DEVELOPMENTS IN PLASTICS TECHNOLOGY—1

Extrusion

THE DEVELOPMENTS SERIES

Developments in many fields of science and technology occur at such a pace that frequently there is a long delay before information about them becomes available and usually it is inconveniently scattered among several journals.

Developments Series books overcome these disadvantages by bringing together within one cover papers dealing with the latest trends and developments in a specific field of study and publishing them within *six months* of their being written.

Many subjects are covered by the series including food science and technology, polymer science, civil and public health engineering, pressure vessels, composite materials, concrete, building science, petroleum technology, geology, etc.

Information on other titles in the series will gladly be sent on application to the publisher.

DEVELOPMENTS IN PLASTICS TECHNOLOGY—1

Extrusion

Edited by

A. WHELAN and D. J. DUNNING

London School of Polymer Technology,
The Polytechnic of North London, Holloway, London, UK

APPLIED SCIENCE PUBLISHERS
LONDON and NEW YORK

APPLIED SCIENCE PUBLISHERS LTD
Ripple Road, Barking, Essex, England

Sole Distributor in the USA and Canada
ELSEVIER SCIENCE PUBLISHING CO., INC.
52, Vanderbilt Avenue, New York, NY 10017, USA

British Library Cataloguing in Publication Data

Developments in plastics technology.—1—(The Developments series)
1. Plastics—Periodicals
I. Series
668.4'05 TP1101

ISBN-13: 978-94-009-6624-6 e-ISBN-13: 978-94-009-6622-2
DOI: 10.1007/978-94-009-6622-2

WITH 21 TABLES AND 87 ILLUSTRATIONS

Softcover reprint of the hardcover 1st edition 1982

Photoset by Interprint Ltd. Malta

PREFACE

In the field of plastics technology, the process of extrusion is widespread and important. It is employed in the compounding and pelletising of plastics materials, in their conversion into products (such as profiles, pipe, hose, sheet, film or bottles) and in the coating of wires, cables, paper, board or foil. A major reason for its use is the screw extruder's ability to melt efficiently and pump continuously large amounts of plastics materials. The understanding of the melting/pumping operation of the extruder and the development of larger and faster-running machines so as to give higher outputs have been given great attention and the results have been widely published.

However, the whole manufacturing technology for extruded products has also developed, particularly in recent years. This has occurred not only by the use of modern screw extruders, but also by the incorporation of improved process control systems, the better design of dies and extrudate handling machinery and by the utilisation of improved plastics materials and additives. It is the purpose of this book to present selected topics which contribute to, or exemplify, these developments in extrusion-based processes.

Some of the chapters have been written by colleagues from industry, each of whom is a specialist in a particular field. We would like to express our thanks to them and to their companies for their participation in the preparation of this book. Our grateful thanks are also given to the Governors of the Polytechnic of North London for allowing us to act as editors and for providing the facilities without which this book could not have been produced.

A. WHELAN
D. J. DUNNING

CONTENTS

Preface v

List of Contributors ix

1. Measurement and Control of Temperature 1
   G. F. Turnbull

2. Extrusion of Cellular Thermoplastics 45
   K. T. Collington

3. Blown Film Production 75
   D. J. Dunning

4. PVC Window Manufacture 155
   R. G. Bruce Mitchell

5. Blow Moulding Processes 195
   V. C. Hind, H. B. Hall and K. Whitehead

6. Rheology and Die Design 229
   R. S. Lenk

Index 281

LIST OF CONTRIBUTORS

K. T. Collington

FBC Ltd, Hauxton, Cambridge CB2 5HU, UK

D. J. Dunning

London School of Polymer Technology, The Polytechnic of North London, Holloway, London N7 8DB, UK

H. B. Hall

Hamilton Machinery Sales Ltd, Sterling House, 20 Station Road, Gerrards Cross, Bucks SL9 8EW, UK

V. C. Hind

Hamilton Machinery Sales Ltd, Sterling House, 20 Station Road, Gerrards Cross, Bucks SL9 8EW, UK

R. S. Lenk

London School of Polymer Technology, The Polytechnic of North London, Holloway, London N7 8DB, UK

R. G. Bruce Mitchell

Glass & Glazing Federation, 6 Mount Row, London W1Y 6DY, UK

G. F. Turnbull

Eurotherm Ltd, Faraday Close, Durrington, Worthing, West Sussex BN13 3PL, UK

K. Whitehead

Hamilton Machinery Sales Ltd, Sterling House, 20 Station Road, Gerrards Cross, Bucks SL9 8EW, UK

Chapter 1

MEASUREMENT AND CONTROL OF TEMPERATURE

G. F. TURNBULL

Eurotherm Ltd, Worthing, UK

SUMMARY

The increasing demands of obtaining greater throughput with increased quality from extrusion products have made accurate measurement and control of temperature more important than ever. This chapter describes in general terms techniques of measuring temperature, heating and cooling systems, and concepts of closed loop control and then contrasts conventional and microprocessor implementations. Since the more exacting requirements demand that process control schemes be applied to extrusion it is not possible to consider the temperature problem in isolation. Therefore a total control system for an extrusion factory is discussed and finally an attempt is made to look into the future.

1. INTRODUCTION

1.1. The General Problem

A general block diagram of the extrusion process is shown in Fig. 1. Here the basic inputs and outputs are shown. There are two basic types of input variable affecting the process and causing changes in outputs:

(a) Controlled or deterministic variables—normally measured directly but sometimes inferred.
(b) Uncontrollable or random variables—not generally measurable on-line.

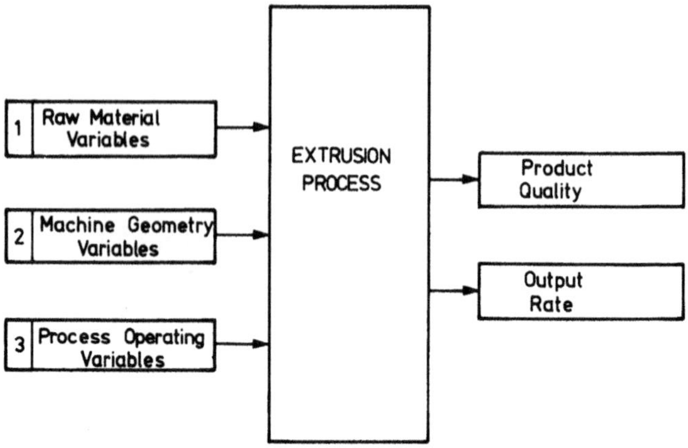

FIG. 1. Extrusion process.

Variables classified in Groups 2 and 3 of Fig. 1 are generally of type (a). Group 2 variables include design of screw, die, heaters, etc; all are under the control of the machine builder, although once the machine is operating they are largely fixed. Here the usual complaint of the systems engineer can often apply, i.e. that insufficient attention to the mechanical design of the machine affects controllability. This point is amplified in later sections. Group 3 variables of Fig. 1 can be subdivided into machine variables (barrel and die temperature profiles, screw speed, pressure, etc.) and process variables (principally melt temperature and pressure, but possibly extending to homogeneity and viscosity).

Group 1 variables include variations in such properties as specific heat, thermal conductivity, friction characteristics and bulk density. Obviously every possible attempt will be made to control these variables but some variation will still occur. The effect on final product quality will depend on several factors, not the least of which is high throughput. This is particularly severe in the area of homogeneity. The conflicting requirements of greater quality and higher throughputs make it essential to control all machine variables very tightly and make control of process variables more desirable. It must be pointed out, however, that the latter requires complex transducers and a good knowledge of machine–control interrelationships and hence should only be contemplated once the tightest control possible on all machine variables still does not give the desired result: this is considered in more detail in Sections 4 and 5. Figure 2 is a specific form of Fig. 1 in which control of the die inlet conditions is attempted with transducers for melt pressure and temperature.

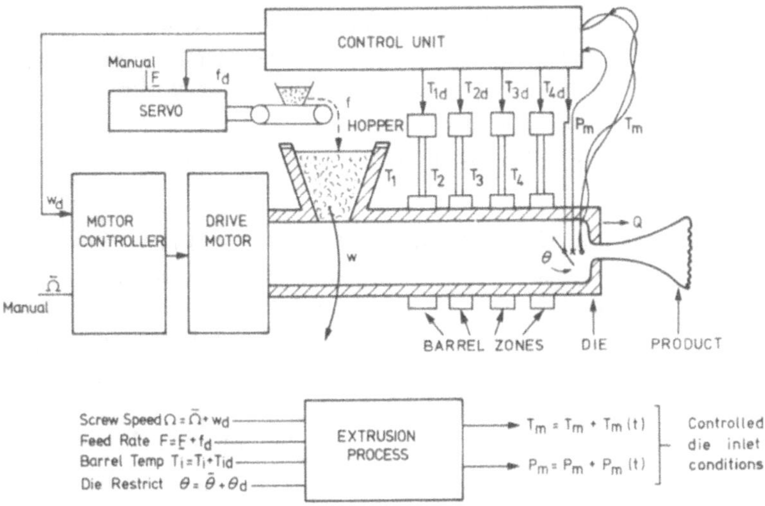

FIG. 2. Extrusion process control.

The input variables in Fig. 2 are screw speed, feed rate, barrel temperatures and die restrictor valve position. Considering steady-state conditions,

$$T_m = \alpha_1 \Omega + \alpha_2 F + \alpha_3 \theta + \sum_{i=1}^{i=n} \kappa_i T_{id}$$

$$P_m = \beta_1 \Omega + \beta_2 F + \beta_3 \theta + \sum_{i=1}^{i=n} \mu_i T_{id}$$

where α, β, κ and μ are gain coefficients and Ω, F, θ, and T_i are as shown in Fig. 2.

In theory, control of melt temperature, T_m, and melt pressure, P_m, could be exercised by varying just two of the die zone temperatures, although too severe a distortion of the profile may have a detrimental effect on, say, homogeneity or surface finish; hence it may be desirable to vary more zones. More seriously, though, rapid pertubations in T_m and P_m cannot be coped with by changing barrel temperatures and require the application of two rapid-action variables—screw speed and die restriction normally being the best choices although feed rate could be used in some cases.

The control scheme (Fig. 2) is facilitated by:

(i) a well insulated melt-temperature thermocouple properly sited;
(ii) a suitable design of melt pressure transducer located close to (i);
(iii) controllers on the barrel zones capable of very tight control and remote setting and supervision;
(iv) a suitable drive system (again capable of remote setting and supervision).

The work heat generated from the action of the screw in mixing and compressing the plastic increases in significance as output rates are increased. This contributes very significantly to the energy required for correct working conditions. The most significant factors affecting the amount of work heat are the type (even the batch) of material and screw speed: cooling is frequently required.

Temperature control aspects of the extruder will now be examined.

1.2. Temperature Control

Temperature control of barrel and die zones is necessary as it is essential to maintain the viscosity of the material constant.

Deviations in viscosity—a complex function of temperature—can cause degradation of material or produce undue stress on the screw and associated drive. As outlined above it is the pressure and temperature of the material as it passes the die which must be kept to very close tolerances if precise dimensional tolerances of the finished product are to be obtained. The first step is accurate control of barrel zone temperatures. Nozzle or die zones normally affect only surface finish, the material being in these zones a relatively short period of time. However, these zones are usually subject to local disturbances so again tight control is necessary.

Before dealing specifically with barrel and die zone control it is necessary to look at the various parts of a temperature control loop, i.e. sensors, final elements and types of control.

2. SENSORS

2.1. Measuring the Temperature

The first item in the control system to consider is the measuring element, of which there are two basic electrical types: active and passive.

The active type are thermocouples. These are formed by the junction of two dissimilar metals and give an output voltage which increases with an increase in the difference in temperature between the thermocouple point of measurement and a reference position. The fact that the millivolt output of the thermocouple in relation to temperature is non-linear and that it depends on a stable reference temperature for comparison purposes are factors which must be taken into account in the controller. Thermocouples are very robust mechanically (this is an obvious advantage in the environment of the moulding shop). They also exhibit good reproducibility from one example to another of the same type. The two most commonly used types in plastics processing are both base metal thermocouples, nickel chrome/nickel aluminium (Type K) and iron/constantan (Type J). Both these couples have a normal tolerance of ±3°C over their working range of 0–400°C, although this can be improved by calibration.

A passive type of transducer is a platinum resistance element. However, this is rarely used on plastics machinery.

2.2. Thermocouple Location

The metal barrel has a heavy cross section so that it can withstand internal pressures. For ease of control it is divided into zones. The most common form of heating is electrical and is applied using band heaters strapped around the barrel (Fig. 3). A controller of any kind can only control the temperature at the point of measurement. Ideally this will be as deep into the barrel wall as possible, since it is the temperature of the

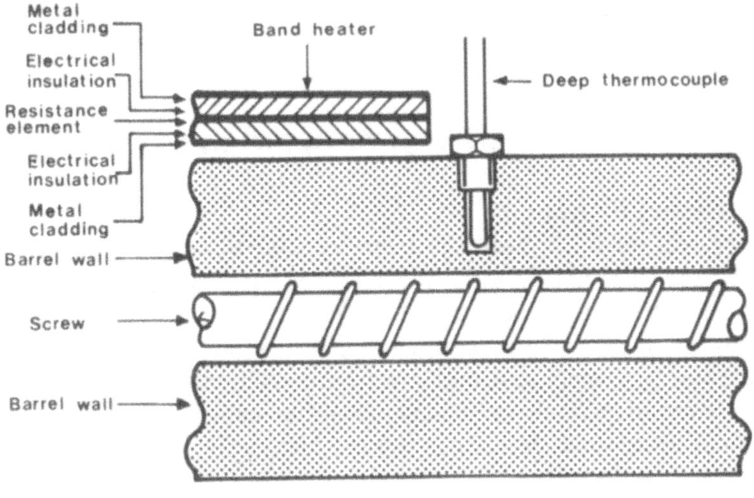

FIG. 3. Section through barrel wall and heater.

plastic which is required and not that of the barrel. Plastic is a poor thermal conductor and depending on whether the net heat flow is into or out of the plastic, a thermocouple deep into the barrel wall will register a temperature above or below the actual temperature. If the measuring element is shallow or on the barrel surface, the difference between the measured and actual melt temperature can be very large. For any given conditions of operation there will be a more-or-less fixed difference between the melt and measured temperatures and acceptable product may be obtained. If, however, the conditions (e.g. machine speed or ambient temperature) change, this temperature difference will change and may give rise to a melt temperature which does not result in an acceptable product. It is therefore important to place the thermocouple as close to the melt as possible (i.e. deep in the barrel) both from steady-state and transient considerations.

2.3. A Combined Deep/Shallow Arrangement

The main problem with a deep couple is the difficulty of achieving good response to loop setpoint changes, this problem being especially acute during start-up where overshoot must be avoided. However, in practice with special overshoot inhibition techniques (mentioned in Section 4) acceptable results are achieved.

An even better response can be achieved with two couples (one shallow and one deep) which are used to implement a cascade control scheme (Fig. 4). The exact details of the control philosophy for this configuration are given in Sections 4.6 and 5.1. It is sufficient at this point to note the shallow couple gives quick response, the deep couple gives accurate measurement and the difference gives power flow in or out of the barrel.

3. HEATING AND COOLING SYSTEMS

3.1. Heating Systems

The correct choice of heater is essential to good control As previously stated, the cheapest form is the strapped-on heater which can give adequate results but unfortunately suffers from relatively short lifetime due to hot-spots and has a rather sluggish response. Greater efficiency and a very fast response is obtained from induction heaters but these are expensive and difficult to replace. A heater capable of giving a good response speed and which has a long lifetime is Pyrotenax cast into aluminium half-shells.

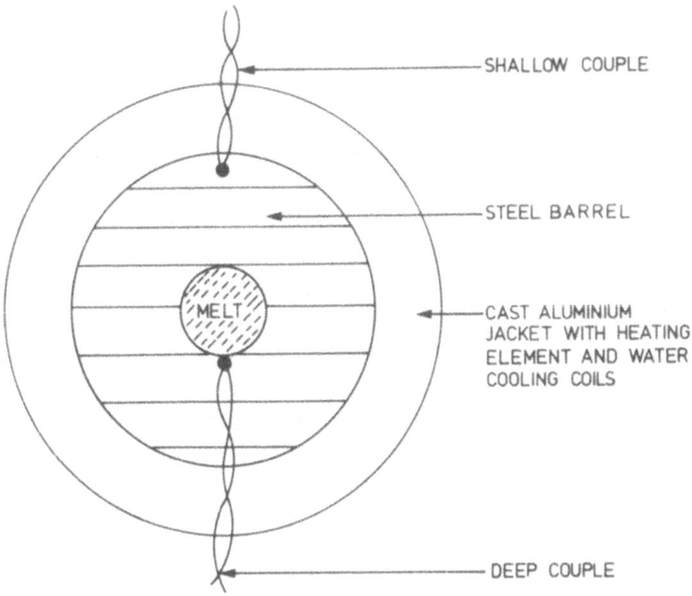

FIG. 4. Deep/shallow cascade system on heat/cool barrel zone.

3.2. Cooling Systems

Higher machine throughputs lead to the requirement for cooling on some barrel zones. The cheapest form of cooling is air-blast, which can be reasonably effective and fast-responding but can achieve only moderate cooling levels even with cooling fins: problems include the necessary provision of ducting, and high noise level. For high cooling rates, fluid cooling is employed. Water is most commonly used and is fed through cooling tubes mounted around the machine barrel, a very good scheme is to cast cooling tubes together with Pyrotenax heaters into the same aluminium half-shells. This ensures that both the heating and cooling systems have the same (fast) response. On-line use of mains water gives scaling problems so that where possible closed-circuit cooling should be employed. Short bursts of water flashing to steam are most efficient.

4. CONTROL CONSIDERATIONS

4.1. On/Off Control

The simplest form of controller provides ON/OFF control of load power. The measured temperature is compared with the set-point and if it is too low, power is applied to the load; if it is too high the power is switched off. In practice there will be a small amount of hysteresis in the controller (mainly so that spurious noise signals on the thermocouple and effects due to mains regulations should not result in rapid ON/OFF chattering of the load power control relay).

If the thermocouple and heater are in very close proximity, i.e. there is no appreciable lag, the temperature will cycle with an amplitude somewhat in excess of the controller hysteresis and with the natural period of the system. There will inevitably be some overshoot on start-up because full power will be applied to the load until the set and actual temperatures become equal and any stored energy in the heater will continue to be transferred to the load even after switch-off. It can be seen that if the thermocouple is deep in the barrel (thus measuring the melt temperature more closely) the system lags will be considerably increased and the temperature cycling will be of a longer period and attain greater amplitude. Similar comments apply to the start-up overshoot.

Thus, in the least demanding circumstances, an ON/OFF controller with a shallow thermocouple may give acceptable results. However, with the large heaters required to give short start-up, overshoot will probably be unacceptable for all but the least demanding situations and will be worse if the thermocouple is sited correctly (i.e. deep within the barrel).

The natural period of the system results from a combination of heater power, heater location, sensor location, and the thermal mass of the system. Performance of an ON/OFF controller is shown in Fig. 5.

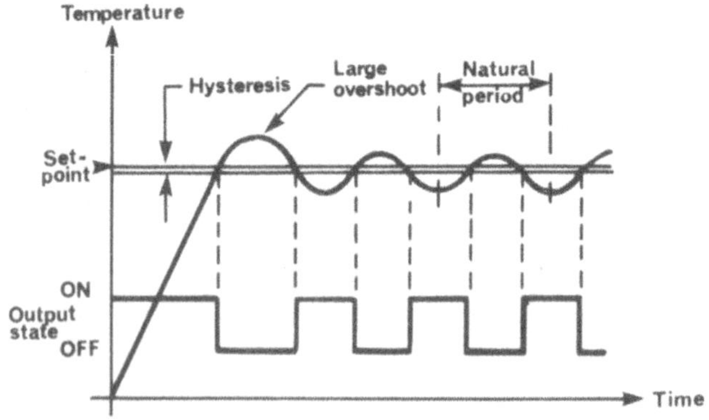

FIG. 5. ON/OFF controller performance showing temperature cycling.

4.2. Proportional Control (P only)

For a proportional controller, the output is switched by the circuitry in the controller at a higher rate than the natural period of the corresponding ON/OFF control loop. Additionally the time-on/time-off ratio (the mark/space ratio) is adjusted by the controller. As the measured temperature approaches the set temperature, the relay will switch off (for a short time) the power supplied to the load. This point, at which just less than full power is applied to the load, is the lower edge of the 'proportional band'. As the actual temperature approaches the set temperature more closely, less and less power is applied to the load until, when the two become equal, the power input is zero. It is general for the proportional band to be downscale of the set-point, i.e. at set-point the power fed to the load is zero.

The proportional band is usually defined as a percentage of the controller set-point scale span. Since the power applied to the load is proportional to the error or difference between actual and measured temperature (a so-called error-actuated system), it follows that if any power is required to maintain the temperature there must be some error in the system. This error is known as offset or droop (Fig. 6). Since on

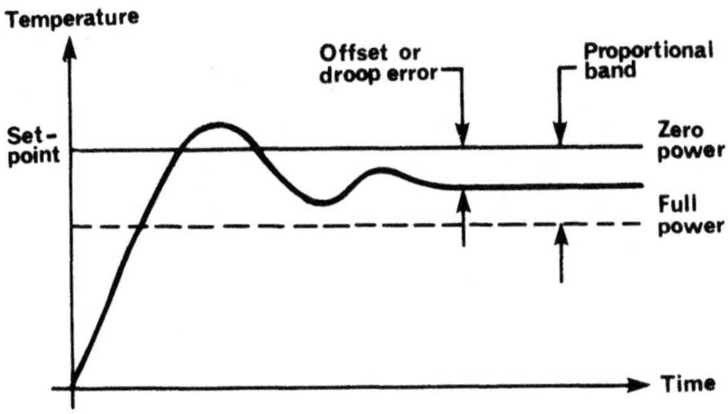

FIG. 6. Proportional control. The proportional band is wide and there is a small overshoot.

start-up the load power will first be switched off at a temperature below the set-point, the resultant overshoot will be reduced compared with an ON/OFF system. With a sufficiently large proportional band and sufficiently rapid cycling of the output power (compared with the natural frequency of the system) the oscillations in temperature will cease eventually. This does not necessarily mean that there will be no start-up overshoot in temperature, but only that the subsequent oscillation will decay to zero amplitude.

Further increases in width of the proportional band beyond that width just resulting in stability (i.e. cessation of a steady-state oscillation) do in fact reduce the initial start-up overshoot magnitude as well as making the unwanted oscillations die away more quickly. As the width of the proportional band is increased the initial start-up overshoot and temperature oscillations will tend to disappear.

To have a significant effect on overshoot behaviour and subsequent oscillation (sufficient to stabilise the system against continuous oscillation in temperature) the proportional band must have a magnitude or width of the same order as the peak-to-peak oscillation magnitude of the ON/OFF system.

Because the ON/OFF cycling time in the proportioning system is made (in the controller design) to be shorter than the shortest oscillation period normally encountered, the fluctuation (not oscillation) in temperature (which is caused by the proportioning ON/OFF action) is very much smaller than the oscillation in an ON/OFF system.

Since the output of the controller is proportional to error, we can assume that for a particular set of machine conditions, and thus for a particular power input requirement, a particular error will result. If the proportional band is wide enough to improve significantly on the start-up performance, the error may be unacceptably large. This can be overcome in part by the addition of an 'offset' control (known as manual reset) with which the proportional band is moved relative to the set-point. However, any change in machine conditions will result in a different power input requirement and a new control point. Additionally, it can be seen that with proportional control the machine temperature will not be held very precisely unless some account is taken of changes in supply voltage (see Section 4.5).

4.3. Proportional plus Derivative Control (PD)

The initial overshoot in a proportional-only control system could probably be overcome by a sufficient widening of the proportional band but the control would then be too 'sloppy'. A significant reduction in overshoot can be achieved (while still maintaining a proportional band narrow enough to give stable control) by adding to the controller a 'derivative' term (*D*).

A differentiating circuit is included and its output, which has a magnitude proportional to the rate of change of the temperature, is added to the controller signal which is proportional to the error. It is added in such a way as to try to minimise the rate of change. For example, the temperature may be below the set-point and, if it is rising rapidly, derivative action would cause the load power to be reduced to a value less than the proportional output corresponding to that particular error. If the derivative term is chosen to be of the right magnitude, i.e. sufficient to counteract the lags in the system, overshoot can be minimised or even eliminated (Fig. 7).

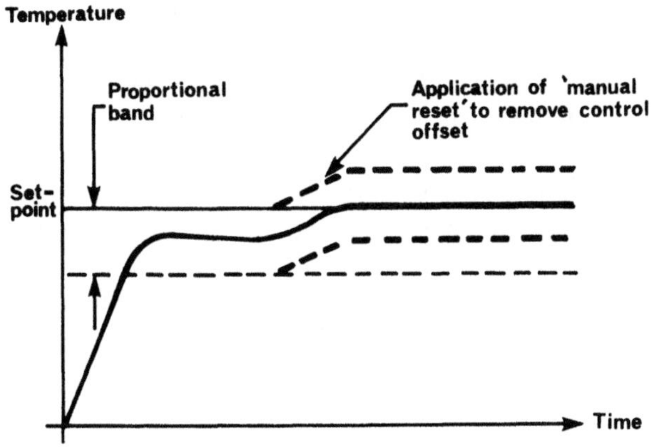

FIG. 7. PD control showing no overshoot and the effect of manual reset.

With an error signal that changes linearly with time, the differentiator output will be of constant magnitude: the time for the error signal then to change through an amount equal to the (constant) output from the differentiator is defined as the derivative time constant.

Derivative action obviously has a beneficial effect during normal machine operation since the controller reaction to any kind of disturbance (e.g. machine starting or stopping) will be much stronger, resulting in tighter control. Proportional plus derivative control may permit the use of deep thermocouples and a proportional band narrow enough to give tolerable control for many undemanding situations.

4.4. Proportional plus Integral plus Derivative Control (Three-Term or PID)

The accuracy limitations of a P or PD controller (as described in Sections 4.2 and 4.3) are due to the fact that an error is necessary for power to be applied to the load. The error will vary, depending on the load power requirements, but for steady conditions can be adjusted out by moving the proportional band relative to the set-point (manual reset).

This cannot be considered a very satisfactory approach since there are many external influences on the machine which will cause the error to change.

A means of automatically removing the steady-state error is to include in the controller an integrator. This slowly accumulates a small error signal so as to give, eventually, from this small error a larger power-control signal which is applied in such a direction (or sense) as to reduce the error. The action of the integrator output provides a bias signal to the power-output stage similar to that provided by the manual reset adjustment mentioned earlier. The proportional error acts in the same sense and this integral of error simply adds to the output power thus applied to the load.

In this way, it is not necessary for a steady-state error to exist (to supply the control signal needed for a given required mean level of output power) because the required power-control signal is now the output signal of the error integrator, which will remain at its fixed level even if the input is zero—i.e. zero error (Fig. 8).

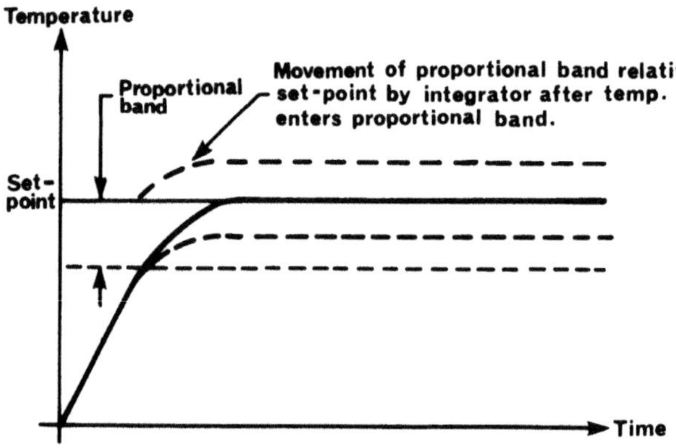

FIG. 8. PID control showing no overshoot and automatic control at set-point.

The integral action is 'sluggish', so that some time must elapse before the error 'integrates out'. The integration results in zero bandwidth because the steady-state gain of an integrator is very large, and yet it does not cause the system to oscillate continuously (as would the zero bandwidth ON/OFF controller) because of its sluggish response. In fact, the integral term is a very definite destabilising influence, and must be made sufficiently slow-acting if it is not to be troublesome in this way.

The integral control term is destabilising in nature because it remembers or accumulates the error which has been occurring, and therefore causes the power-output to switch off at a later instant than otherwise: how much later (and therefore how much destabilising effect this has) depends upon the magnitude of the integral signal—i.e. upon the integral action time or time constant. This is defined as the time, with a given constant error-signal, for the integral control signal to become equal in magnitude to the given proportional control signal. Larger integral time constants cause fewer destabilising effects and the integral time constant must in any case be longer than the natural period of the corresponding ON/OFF controller for severe destabilising effects not to occur. For this reason PI controllers are not generally applied to plastics processing equipment, but rather PID control is used as this gives extra stability.

In order to minimise the 'storage' of unwanted error signals (e.g. those occurring when the error is very large shortly after switch-on) it is necessary to include as part of the 'overshoot' inhibition system some means of removing the integrator from the circuit until the error is relatively small. This is usually effected by 'shorting out' the capacitor on which the integral information is stored if an output power greater than 100% is being demanded (i.e. if the measured temperature is below the proportional band).

Some control equipment is described as PDPI rather than PID. (PDPI is a special implementation of PID where a separate derivative circuit is placed ahead of the normal 3-term control circuit.) However, examination of the transfer functions of these two circuit configurations usually reveals that there is in practice no significant difference between the two and indeed the double derivative of the PDPI can lead to inferior noise performance.

4.5. Available Control Equipment

Controllers used on extruders can be, in the simplest case, ON/OFF with no indication of temperature, or PD and PID with indication. ON/OFF and PD controllers are generally used in low-cost applications where their performance may be adequate. PID control is required in instances where a high stability of temperature must be maintained under all operating conditions (Fig. 9).

Controllers may be fitted with an analogue set-point mechanism (preferably of the drum-scale type, to provide good scale length for accurate temperature setting and reading) or with a thumb-wheel digit switch and error meter. The latter arrangement is particularly useful when

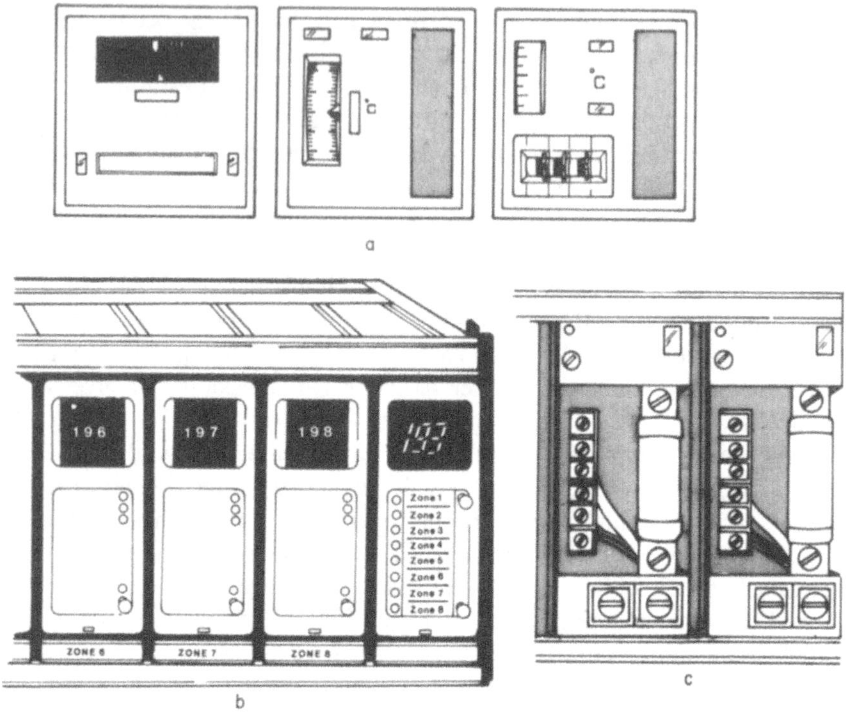

FIG. 9. A range of PD and PID controllers as (a) single units and (b) in a multiple racking configuration with shared digital indicator, together with (c) thyristor units rated up to 50A (480V) for mounting inside a cubicle.

set-point repeatability is of importance and also in avoiding interpolation errors due to misreading of scale graduations. Typical set-point accuracies are 0·5% of set-point span, i.e. ± 2°C for a typical 0–400°C controller scale. This is at least as good as the differences found between different examples of the same type of thermocouple. Set-point accuracy is in any case not as important as stability and repeatability since often the required set-point is arrived at experimentally but once obtained it must then be maintained accurately.

Cold junction compensation, usually in the form of an electronic reference, is built into the controller and should achieve a 20 to 1 rejection of ambient temperature changes, i.e. the load temperature alters not more than 0·5°C for every 10°C change in ambient temperature. The mains voltage can change quite substantially in an industrial environment and a temperature controller should reject changes of -15 to $+10$% of the nominal supply voltage.

The controller output stage can be fitted with power feedback. If a measure of supply voltage is used to determine the ratio of load power ON to OFF times for a given control signal within the proportional band, then disturbances which would otherwise occur are minimised. This feature is particularly important since power is proportional to the square of the applied voltage and hence a 5% change in supply voltage will cause a 10% change in load power, which may give rise to a considerable disturbance in the load temperature. Power feedback may reduce it to as little as 1 or 2%, leaving only this small disturbance to be removed by the temperature control loop.

Controller outputs are generally of two types: (i) relay or triac, or (ii) thyristors. Relays and triacs are usually used with external contactors for an economical power switching system. The rate of switching is a limitation and would be chosen to be fairly slow so as to avoid excessive wear on the moving parts. However, if the rate approaches the natural period of the system, significant temperature ripple will result. A compromise has to be found which generally results in typical (ON + OFF) times of 20 s or so. A contactor has an expected life of around one million operations; thus, it would need to be replaced after about $2\frac{1}{2}$ years if the machine was in use for 8 h each day.

Thyristors, however, because of their 'solid-state' nature, can switch the load power very rapidly without any wear and can therefore offer long life and better control of temperature. They can be operated in one of two ways: phase angle and fast cycle. In the case of phase angle, the thyristors are switched on in every half cycle of the mains supply and the load power is controlled by varying the point in the cycle at which switch-on occurs (Fig. 10).

More commonly on plastics machinery, fast cycle firing (burst firing) is used. Here the thyristors are switched on at the beginning of a half cycle, when the supply voltage is zero, and remain conducting through several cycles, the power input to the load being regulated by the ratio of ON to OFF time. Fast cycling has the advantage of producing little radio frequency interference or supply waveform distortion. Typical (ON + OFF) times are about 0·2 s and this results in very smooth control of temperature. Such rapid cycling minimises thermal stress experienced by the heaters, thus prolonging their life. (Note: even such rapid cycling would be too slow for a lamp-type heater, where phase angle firing would be essential.)

The combination of controller parameters (i.e. the proportional band employed, the integral and derivative times used) together with the

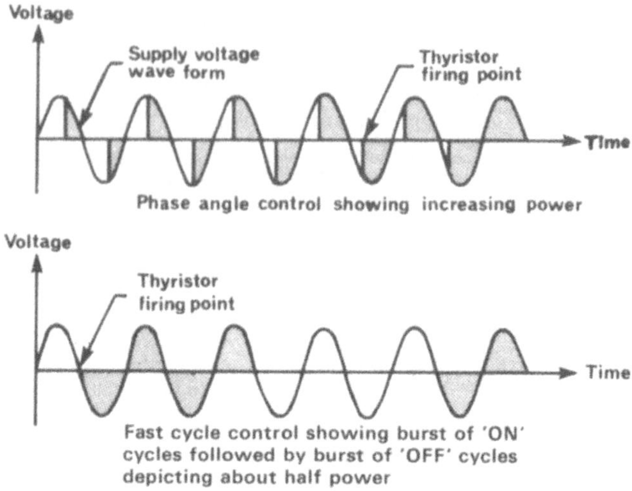

FIG. 10. Phase angle and fast cycle.

process being controlled determines whether the process is stable and whether good start-up performance is obtained. In the case of an ON/OFF controller, the overall performance is determined completely by the process, i.e. the power of the heater and the siting of the thermocouple. A shallow thermocouple may give better results than a deep one since it would reduce the heat transfer lag.

In the case of a PD controller, the proportional bandwidth and the derivative time are of importance. The proportional band would be a compromise between a small value, which may lead to instability and start-up overshoot problems, and a large value, which would give 'sloppy' control. The derivative time is usually fixed at about 1 min. This will give satisfactory performance with most machines, which will, of course, be within the limitations of this type of controller.

It should be noted that since the derivative action has a stabilising effect against system heat transfer lags, if the lag is very small (i.e. with a shallow thermocouple) instability may result. It is therefore advisable to use PD controllers only with deep thermocouples.

When using a three-term or PID controller it should be possible to achieve good start-up performance, good response around the normal working point, and steady-state error, i.e. the PID controller should always control at set-point. Such controllers leave the factory with preset values of proportional band, integral and derivative time (typically 5%, 6 min and 1 min, respectively) and these will give good performance on most machines (for zones with natural periods of less than 6 min). In general the response will be slightly overdamped so as to give some margin of stability. A small sacrifice will therefore be made in that the time for the machine to reach stable operating temperature will be a little longer than the absolute minimum.

There are, however, cases where the 'standard' controller parameters are quite unsuitable and result in intolerable start-up performance or instability. Optimisation of settings is dealt with in more detail in Section 4.7. In Section 5 more details are given of the controller requirements for extrusion.

4.6. Cascade Control

Cascade control as applied to the deep/shallow arrangement of Fig. 4 is shown in Fig. 11. Essentially the output of one loop—the master controller—controls the set-point of a second loop—the slave controller.

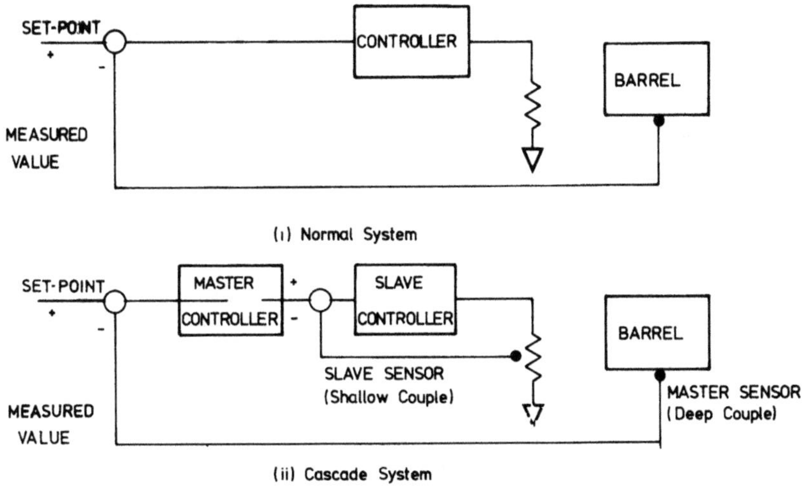

FIG. 11. Cascade control of barrel zone by deep and shallow couples.

The basic advantages of this technique are:

(a) Tighter and more responsive control of melt temperature.
(b) Fast correction via the slave loop of disturbances originating near the slave sensor (for example heater supply variations, ambient temperature variations) resulting in reduced fluctuations at the master sensor from these causes.

(c) Ability to provide safe operating limits on heater temperature by set-point limits on the slave loop.
(d) Easier loop tuning resulting from separating the problem into two parts, the master loop—normally implemented using PI control—and the slave loop—employing PD control.
(e) Interpolation techniques can be used to estimate melt temperature.
(f) Temperature difference deep to shallow gives power flow and facilitates power feedback on both heating and cooling: see Section 5.1.

Previously implementation of the above scheme was very expensive using discrete controllers. However, microprocessor techniques now allow a reasonably priced solution: see Section 5.4 and 6.2.

4.7. Modelling

A fairly simplified picture of closed loop control has so far been described which goes a long way in explaining extrusion. However, for full understanding a greater depth of knowledge is required. Here the use of models (electrical analogies, digital simulation, etc.) is very useful. The models are used by the manufacturer of a temperature controller to make sure his instrument will be suitable for different forms of processing plant. However, the extrusion processor is unlikely to use them except possibly to determine best settings for the controller.

In Section 4.7.1 some basic principles of a single closed loop system will be examined; in Section 4.7.2 the more general problem of a multivariable system will be examined.

4.7.1. Three Differing Systems

A basic closed loop system is shown in Fig. 12(a). Normally three-term control is used for the controller, the purpose of integral action being to integrate out the steady-state error of a proportional-only system while derivative action is to facilitate smaller bandwidths and hence give tighter control than is possible with a proportional-plus-integral system. These are general statements and it is necessary to look at some specific plant transfer functions to see what is required in practice. It has been found useful to look at three quite different categories of plant transfer function.

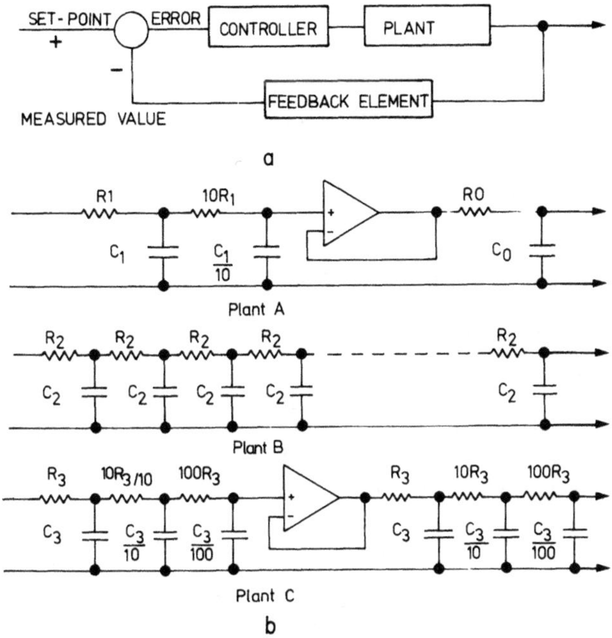

FIG. 12. Basic principles of closed loop control: (a) simple closed loop; (b) electrical analogies of three types of plant—A, narrow bandwidth; B, medium bandwidth; C, wide bandwidth.

These are:

(i) narrow bandwidth —plant A;
(ii) medium bandwidth—plant B;
(iii) wide bandwidth —plant C.

Generally speaking the narrowest bandwidth is possible with a single dominant lag; the widest bandwidth is needed where the plant transfer function contains a number of lags each having the same time constants.

Figure 12(b) shows electrical analogies of the three plants.

The models can be expressed in mathematical equations in which the symbols are defined as follows:

s = Laplace operator
T_n = Time constants
t = Time
P = Proportional band

I = Integral time
D = Derivative time
A, B, C refer to different extrusion plants
f_n = step responses as a function of time

(*i*) *Narrow bandwidth system:*
Plant A has a transfer function

$$A(s) = \frac{1 \cdot 5}{(1 + sT_0)(1 + sT_1)^2}$$

Step response ($T_0 \gg T_1$) is

$$f_A(t) = 1 \cdot 5 \left[(1 - e^{-t/T_0}) - \frac{T_1}{T_0}(2 - 2e^{-t/T_1} - te^{-t/T_1}) \right]$$

Optimum control terms are approximately

$$P = 150\, T_1/T_0,\ I = 8T_1,\ D = 2T_1$$

For example, if $T_0 = 75T_1$, then $P = 2\%$ and I is about one-tenth of the 100% rise-time of the measured value; D is $I/4$.

(*ii*) *Medium bandwidth system:*
Plant B has the transfer function

$$B(s) = \frac{1 \cdot 5}{e^{1/2}(sT_2)}$$

Step response is

$$f_B(t) = 1 \cdot 5 \operatorname{erfc} \left[\frac{(T_2)}{4_t} \right]^{\frac{1}{2}}$$

where

$$\operatorname{erfc}(y) = 1 - \operatorname{erf}(y) = \frac{2}{\sqrt{\pi}} \int_y^{\infty} e^{-u^2} du$$

Optimum control terms will be approximately

$$P = 15\%,\ I = \frac{6T_2}{\pi^2},\ D = \frac{T_2}{\pi^2}$$

Relating again integral time constant to step response, I is about one-third of the 100% rise-time of the measured value; D is $I/6$.

(*iii*) *Wide bandwidth system:*

Plant C has the transfer function

$$C(s)=\frac{1\cdot5}{(1+sT_3)^6}$$

Step response is

$$f_C(t)=\frac{1\cdot5}{5}\int_0^t\left(\frac{t}{T_3}\right)^5 e^{-t/T}dt$$

The solution of this integral is

$$1\cdot5\left[1-e^{-x}\left(1+x+\frac{x^2}{2!}+\frac{x^3}{3!}+\ldots\right)\right]$$

where $x=\dfrac{t}{T_3}$

In this situation derivative action is a destabilising influence and optimum terms are approximately

$$P=250\%;\ I=4\cdot5\ T_3$$

The relation between integral and step response is not of the same relevance as for plants A and B due to non-linear behaviour at such a wide bandwidth.

The relevance of the above to the extrusion machine will be examined in section 5.

4.7.2. The Complete Problem

Now consider the complete problem. Figure 13 shows various parts of a control system, A, B, C and D. With overall feedback (i) to define the process, the closed loop response speed, accuracy and reduction of system noise are restricted by the fundamental requirement of loop stability. With local feedback of A, B, C and D better results in all these respects can be achieved. More generally—as obviously this picture is an oversimplification—it can be stated as a system philosophy that the best approach to a large system is to achieve control of all appropriate variables individually rather than to attempt a sophisticated multiple-loop feedback from only the output quantity.

Applying this philosophy to the problem of the extruder as outlined in the previous section, the following statements can be made:

(i) All barrel (and die) zones must be controlled as tightly as possible; this is dealt with in detail in section 5.1.

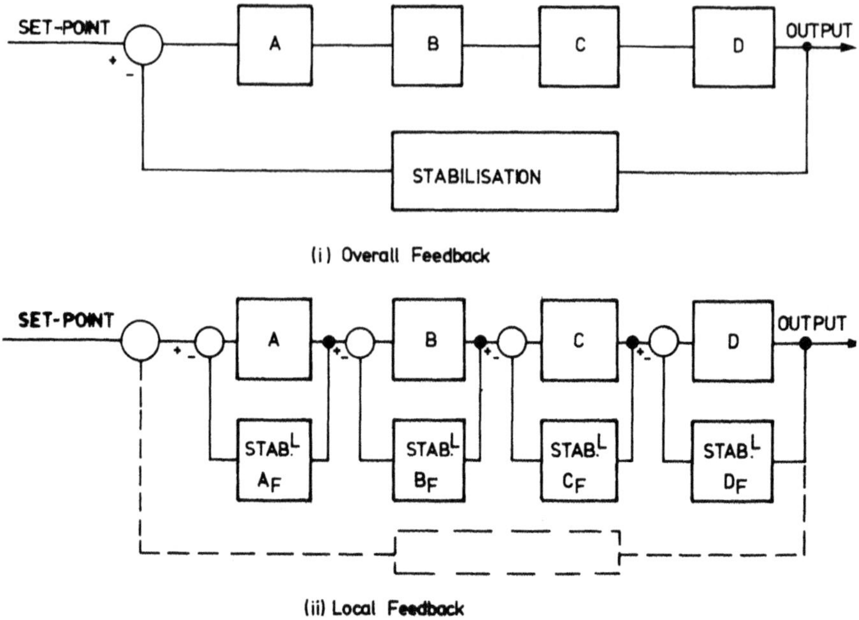

FIG. 13. Control philosophy.

(ii) Having done this, it may still be necessary to control melt temperature (and pressure); this also is discussed later (section 5.2).

(iii) Control of screw speed is necessary.

The cost and complexity of controlling each variable must be weighed against its contribution to total error in the controlled quantity.

5. TEMPERATURE CONTROL OF AN EXTRUDER

5.1. Control of Barrel and Die Zones

5.1.1. An Electrical Analogy

An approximate electrical analogy of a section of barrel is shown in Fig. 14, where current is analogous to power and voltage to temperature. The electrical resistances represent thermal resistances in the extruder. Because of the electrical capacitances, voltage changes lag behind current, i.e. in the extruder the temperature changes lag behind the power flow.

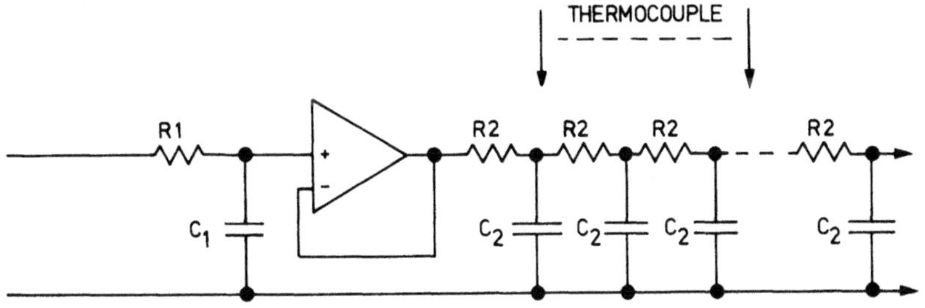

FIG. 14. Electrical analogy of an extruder barrel.

There are essentially two components to the model—a single lag representing the heater and a second component representing the barrel (a further lag is introduced if the heater is not in intimate contact with the barrel). The second component is equivalent to plant B in Fig. 12(b). The thermocouple will be sited somewhere along this line, depending on its depth, and therefore it can be seen that the transfer function is a combination of plants A and B in the previous section.

Figure 4 shows a deep/shallow cascade scheme with the above-described heat–cool implementation.

5.1.2. Control Characteristics

With a well designed heat–cool system (such as that outlined in Section 3) relative heat–cool bandwidths would be set according to relative heating and cooling powers. Different time constants for heating and cooling are possible (especially with a microprocessor implementation) but the need to do this would indicate deficiencies in mechanical design—as would also be the case if bandwidths were not in proportion to power capacities. A good example of this is the use of a crude fan for air cooling, whereby stiction and effective transport lag can give rise to derivative-caused oscillations.

The control characteristics required for both heating and cooling regions are:

(i) properly designed overshoot inhibition;
(ii) well implemented and carefully optimised control.

These aspects have been covered in previous sections.

5.1.3. Output/driver stages

On the heating side, thyristors as control elements will undoubtedly give optimum results with long life and low ripple due to their fast switching.

To obtain a linear relationship for heating requires power feedback in the driver stage: the effects of mains variations can then be eliminated in the driver.

Some possible cooling stages are:

(a) pulsing drive to solenoids: two modes of control are possible—constant pulse width or constant repetition rate. In practice, non-linearity between the control signal and cooling is compensated for in the controller and the resulting control is a hybrid between the two modes, i.e. constant repetition rate at low powers going to constant pulse width at high powers;

(b) valve motor positioner drive for continuous water flow or air flow systems;

(c) ON/OFF proportioning or continuous speed control to an air blast fan. Here again non-linearity must be compensated for, and cooling power is limited.

Power feedback to the cooling stage is more difficult than heat power feedback: it involves sensing power flow in the water or air using thermocouple or thermistor sensors. In this respect the scheme of Fig. 15 is particularly interesting as it allows a measurement of heat flow from the differences sensed by shallow and deep couples. This would facilitate, for example, measurement, memory and subsequent compensation (for non-linear behaviour) with a microprocessor system.

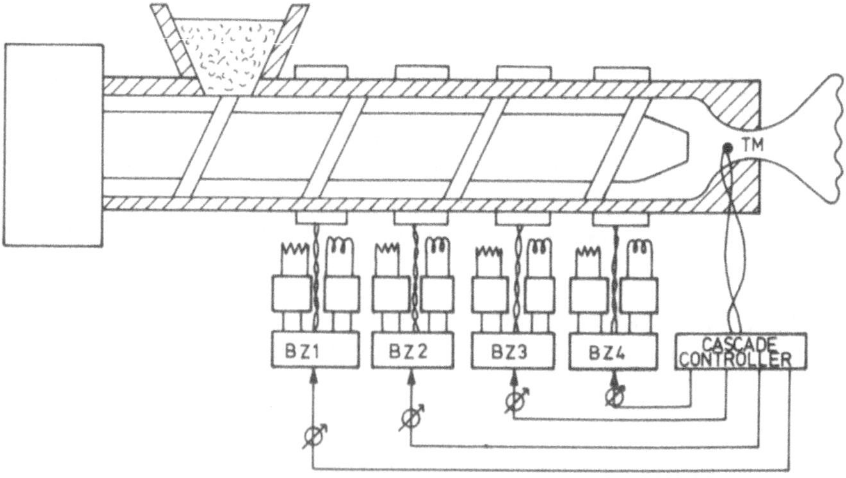

FIG. 15. Melt temperature control.

5.1.4. Alarm and Protection

Other features which can be incorporated in barrel/die zone control are connected with alarm/protection philosophies. Just two examples are:

(a) detection and appropriate action for broken sensors and power elements; and
(b) detection of loose clamping-bolts causing an airgap with an inherent deterioration in control.

5.2. Melt Temperature Control

Melt control on an extruder is shown in Fig. 15. As can be seen this is rather more complex than the simplified picture of Fig. 13 in that both serial and parallel paths are involved. However, the principles outlined apply, specifically because tight control of the barrel zones is an essential first step to close control of melt temperature. The principles of achieving tight barrel zone control have been discussed in Section 5.1.

The implementation of a melt control loop requires firstly an accurate, reliable melt thermocouple with good dynamic characteristics. Disturbance of the plastic flow must also be minimal and cost should be compatible with required performance. Steady-state errors will occur due to heat conduction (down the thermocouple stem), shear heating and compression heating. Dynamic errors are due to probe thermal inertia relative to the surface area it presents to the plastic. Parallel-to-flow probes of adequate fast response and low conduction errors have been developed which have sufficient mechanical strength for most extruders.

The cascade controller differs from the barrel/die controllers because of bumpless transfer (i.e. when a cascade switch is opened or closed there is no change in the controlled temperature). The outer melt loop has to be broken and closed with no bumps in the local controllers. This requires output preconditioning. Bumpless transfer will considerably reduce the settling time problems.

Since there is effectively a transport lag in the outer melt loop, the situation approaches a class C plant. This requires therefore a PI control, D being a destabilising effect. In practice bandwidths around 20% are possible. Additionally, the permitted change in the slave temperature setting is normally restricted to a maximum of 20–50°C.

5.3. Discrete Controller Implementation

5.3.1. Important Features

A block diagram of a barrel/die zone controller is shown in Fig. 16 and some

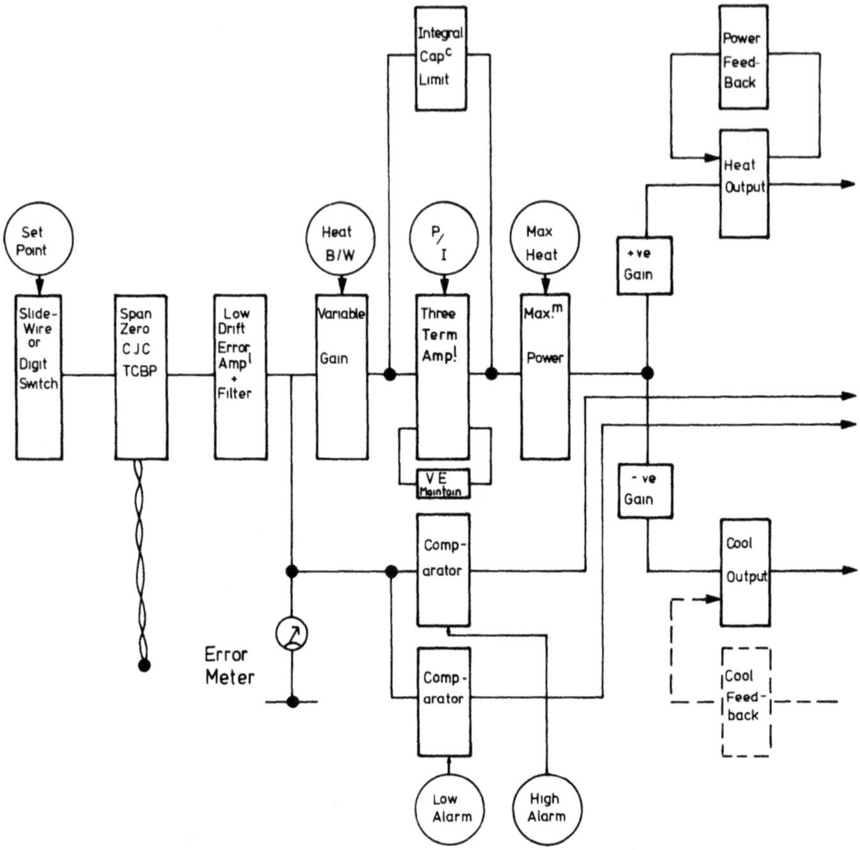

FIG. 16. Discrete barrel zone controller. CJC: Cold junction compensation. TCBP: Thermocouple break protection.

important features are:

(a) Proportioning action. This is necessary because with deep couples and ON/OFF control, oscillations would be unacceptable: integral action maintains control temperature independent of power level; derivative action allows tight control against transients.
(b) Overshoot/undershoot. This inhibition gives fast start-up with no overshoot and good step response. This is achieved using a technique to provide saturation of an operational amplifier (i.e. virtual earth (VE) maintenance) and limiting the voltage on the integral capacitor: these require a defined error window.

(c) Power feedback. This feature on the heat output stage eliminates the effects of mains variations—another example of a tight local loop.
(d) Safety features. Typical safety features are band alarms plus items outlined above for sensor/power device failure. Full scale alarms would require a full scale amplifier, which is not otherwise required.

The melt cascade controller would have all the above features, plus:

(e) Auto-manual switch with bumpless transfer both ways.
(f) Maximum and minimum output controls. These ensure that the degree of cascade feedback permitted is not high enough to give dangerous increases in melt temperature and also not low enough to cause the plastic to solidify.
(g) PI control with values several times higher than those used on the zone controllers.
(h) It is possible to fit adjustable time constants so as to compensate for appreciable variations in machine speed. Automatic adjustment with machine speed is difficult with discrete units but easily possible with microprocessor implementation.

To complete the melt cascade loop (Fig. 15) requires a ganged isolating unit, so as to convert the control output signal of the cascade controller into a number of scaled and mutually isolated signals for application to the barrel zone controllers. Alternatively cascade control to screw speed or back pressure would be implemented in the same way.

Figure 17 shows a racking system with six barrel zone controllers and one melt cascade controller, a six-way ganged isolator unit and a switchable digital indicator.

5.3.2. Controller specifications

Below is the specification of a controller suitable for a barrel zone:

The controller has a three-digit set-point. Setting or trimming of the set-point by a remote signal is also possible but this must be isolated. Deviation from set-point is indicated by three LEDs in a vertical line. Zero deviation results in steady illumination of the green centre light. A deviation causes one of the outer red lights to flash alternately with the green for a time proportional to the deviation magnitude. The upper

FIG. 17. Discrete melt temperature control system.

light corresponds to temperatures higher than set-point. For deviations greater than 10% of instrument span the appropriate red light remains on.

The control action is proportional to integral and derivative action. Control techniques employing a defined error window ensure overshoot-free start-up behaviour.

There are three possible outputs for the heat channel: (i) internal zero-voltage switched triac—providing time proportioning drive direct to contactors or low power heaters; (ii) a faster proportioning logic output—suitable for driving a range of solid state contactors; (iii) a continuous 0–5V dc or other dc output suitable for large thyristors with integral drivers. Both (i) and (ii) are provided with power feedback to prevent supply voltage changes affecting output power.

An optional cool channel has three possible outputs: internal triac for time proportioning zero voltage-switched drive to either fans or solenoids (non-linear feedback is included and its characteristics are preset to suit either water injection or fan cooling); plus dc and logic outputs ((iii) and (ii)) available on the heat channel.

To take account of cool channels having different ratings from the heat channel, the cool proportional band may be adjusted from 0·4 to 2·5 times the heat proportional band. An optional deviation alarm has separate fail-safe high and low isolated logic outputs. They are suitable for series connection to interface with an alarm collection module.

The alarm set-point is adjustable over the range 1–10% of full span. In the alarm state a separate LED flashes on the instrument.

Table 1 summarises the technical specifications required typically for a temperature controller.

TABLE 1
TYPICAL TECHNICAL SPECIFICATIONS OF A TEMPERATURE CONTROLLER

Inputs		
Thermocouple		
US Standard	ASTM E230 (1972)	
British Standard	BS 4937 (1973)	
German Standard	DIN 43710	
Scale ranges	Large selection	
Thermocouple resistance	200Ω max.	
Thermocouple break protection	Upscale standard	
Thermocouple cold junction compensation	Standard: typically 20:1 rejection	
Resistance thermometer		
British Standard	BS 1904	100Ω at 0°C
German Standard	Pt 100 to DIN 43760	138·5Ω at 100°C
Scale ranges	Large selection	
Two- or three-wire connection	Three-wire to compensate for cable runs up to 20Ω/lead	
Outputs		
Available in either the heat or cool channel	Logic output operating in fast cycling mode for solid-state contactor drive (20mA open circuit collector, max. 10V). Unisolated dc output either 0–5V, 0–10V, 4–20mA or 0–20mA. 2A 240V internal triac operating in the slow cycling zero voltage switching mode (suitable for contactor drive).	
Cool channel only	2A 240V ac internal triac operating in the fast cycle zero voltage switching mode with linearisation to suit water injection or blown air cooling.	
Optional functions		
Deviation alarm	Range adjustable ±1–10% of span; preset at 5%. Separate isolated transistor logic outputs for high and low deviations: maximum rating 1·6mA 25V dc. Common flashing red indicator on front panel for either high or low alarm. High impedance output in alarm condition ensures fail-safe operation.	

TABLE 1—*contd.*

General	
Deviation indicator	Red for deviations >10%, flashing between red and green in range ±10% span, steady green when on control (i.e. within ±1%).
Control action	Three-term action with special overshoot inhibition feature, 5 min integral time, 1 derivative time, heat proportional band will normally be preset at 5%, adjustable in the range 1–10%.
Power supply	230V and 115V taps allowing ±15% on voltage and a frequency range 48–62Hz.
Calibrated accuracy	±½% span
Ambient temperature effects	Amplifier drift <2μV/°C
Ambient temperature	0–50°C

5.4. Microprocessor Implementation

A block diagram of a microprocessor controller is shown in Fig. 18.

This embodies many of the same features as a conventional analogue three-term controller, but the digital implementation does overcome some limitations of its analogue equivalent and also poses different problems.

Using fixed-point arithmetic for a microprocessor three-term algorithm imposes a compromise between resolution and maximum handleable signal levels (equivalent to a saturation limit in analogue systems) which can only be achieved by using 16 bit arithmetic. A natural result of this is the 'error limit' shown in the proportional path of Fig. 18 which should be at least 100% of span so as to provide optimal response for wide bandwidth systems.

The measured value input to the controller is initially processed by a hardware filter before it is sampled at regular intervals of T_s seconds duration and converted to a digital signal. The filter prevents aliasing, which is a phenomenon arising when the frequency of sampling is too high relative to the bandwidth of the signal being sampled. Aliasing produces completely erroneous results. The digital measured value is then filtered again (by a first-order difference filter) before subtraction from set-point to obtain the error signal.

The differential loop uses the filtered measured value as an input rather than the error signal since this avoids a large derivative 'bump' when set-point is changed. The differential term is obtained by subtracting the latest measured value from the previous sample and multiplying

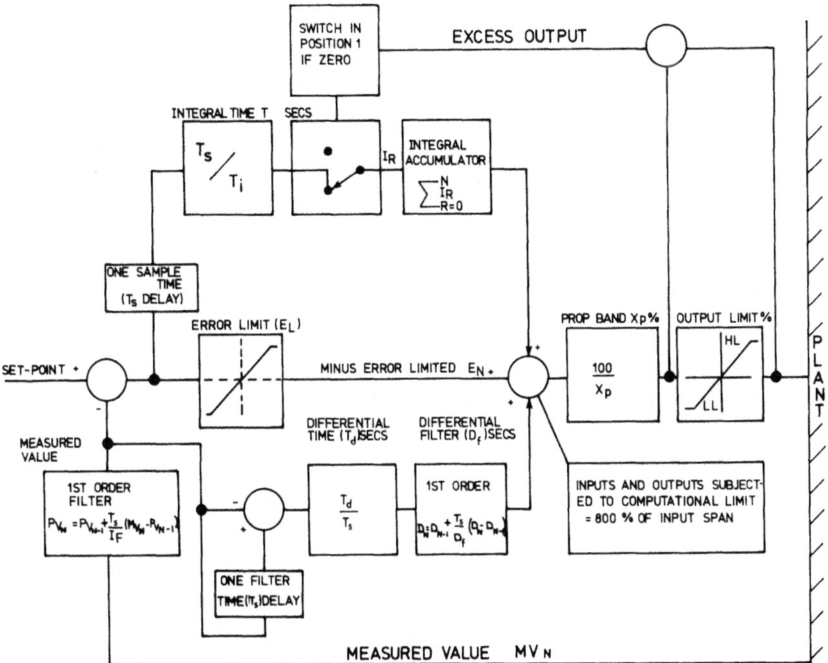

FIG. 18. Three-digital implementation of PID control.

by a scaling factor T_d/T_s where T_d is the derivative time. This differential output is then processed by a further first-order filter with a time constant $D_f = T_d/10$ which effectively limits the saturation problems. As can be seen from Fig. 18, the computational limit at the point where the P, I and D terms are added is 800% of span, allowing for large derivative kicks.

The integral loop uses the error signal as an input and scales this by T_s/T_i where T_i is the integral time. For small signal deviations this scaled error signal is continuously added into the integral accumulator and the output from this accumulator is added to the P and D terms, thus completing the integral loop.

The summed P, I and D terms are then scaled together by a factor $100/X_p$, where X_p is the proportional band as a percentage of the span. This control signal is then compared with output limits which vary between zero and $\pm 100\%$ depending on the maximum power setting.

If the control signal exceeds the output limits the output is set to that limit and the excess output is flagged which breaks the integral loop, thus preventing the integral from 'winding up'. An alternative strategy is to actually wind the integral back to set the proportional band on the appropriate side of set-point in this condition.

For the small signal control situation the output varies within its saturation limits and the loop behaves in a very similar manner to conventional analogue controllers.

6. PROCESS CONTROL IN AN EXTRUDER FACTORY

6.1. The Basic Problem

Restating the problem, the basic need for an extruder is to maximise throughput whilst maintaining acceptable quality and minimise operating costs. The manufacturer of extrusion products has to maintain consistency, reliability and performance from his production system, coupled with competitive levels of productivity, in order to satisfy his customers' demand for high quality products which meet their specification at the lowest possible price. This requires therefore better design of extrusion machines with increasingly sophisticated control systems. Furthermore these control systems must integrate into an overall production control system, i.e. good up-to-date information flow is essential.

The basic principles of levels of control and information flow for a general manufacturing plant are illustrated in Fig. 19. The whole manufacturing system can be co-ordinated by a well designed production control system which has various specialised process control systems as subsystems. This requires up-to-date information available to supervisors as and when they require it in order that they can exert the appropriate controlling actions. Thus the various process control systems, as well as performing the appropriate control tasks, must make available this information on the various variables within the system. Microprocessor-based control systems make the implementation of such a system feasible at an economic price.

In the context of an extrusion factory the general scheme would be as shown in Fig. 20. This shows a system based on distributed intelligence:

(i) Each machine is separately instrumented but communicates with a central computer along a serial communication bus (Fig. 20(a)).
(ii) If the machine is sufficiently complex it will be beneficial to split it into several parts. Figure 20(b) shows a coating line instrumented in this way.

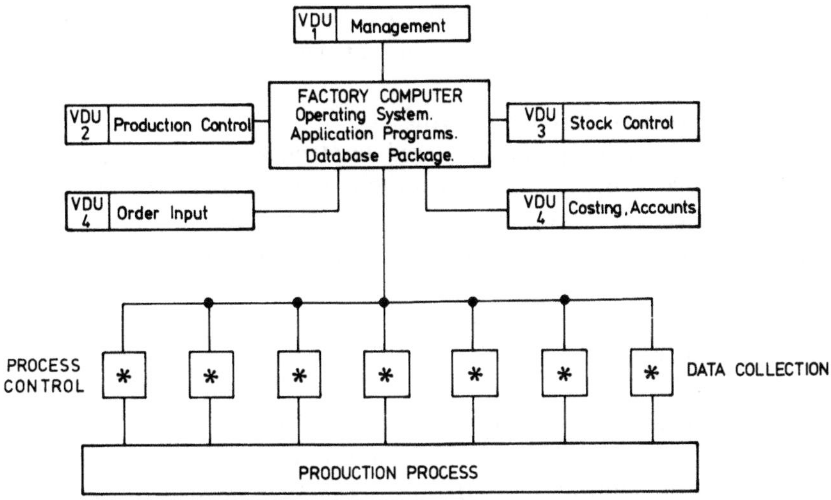

FIG. 19. On-line real-time production control system.

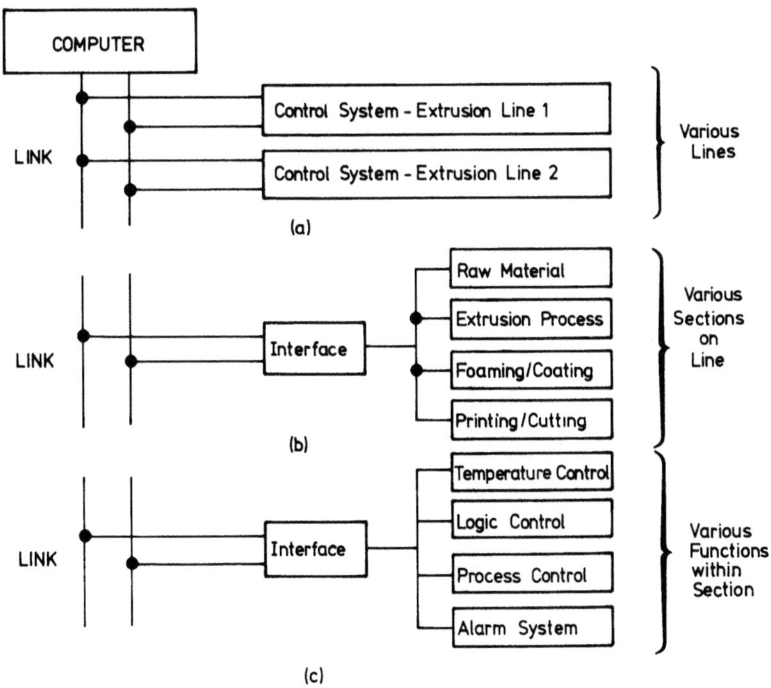

FIG. 20. Control systems in an extruder factory. (a) Overall scheme; (b) large extruder; (c) section of (b).

(iii) Within each section a further split is useful. Figure 20(c) shows instrumentation for each section of the basic extruder.

6.2. Microprocessor Control System

All schemes for microprocessor-based implementation of the extruder system involve a digital implementation of the three-term algorithm, which has been described in section 5.4. There are two fundamental approaches to the problem and a variety of hybrid solutions lying between these.

6.2.1. Fundamental Approaches

6.2.1.1. Direct digital controller (d.d.c.) solution. Such a scheme is illustrated in Fig. 21. This is the conventional minicomputer-based structure for a multi-loop control problem, but updated using microprocessor technology. Each input convertor performs the following

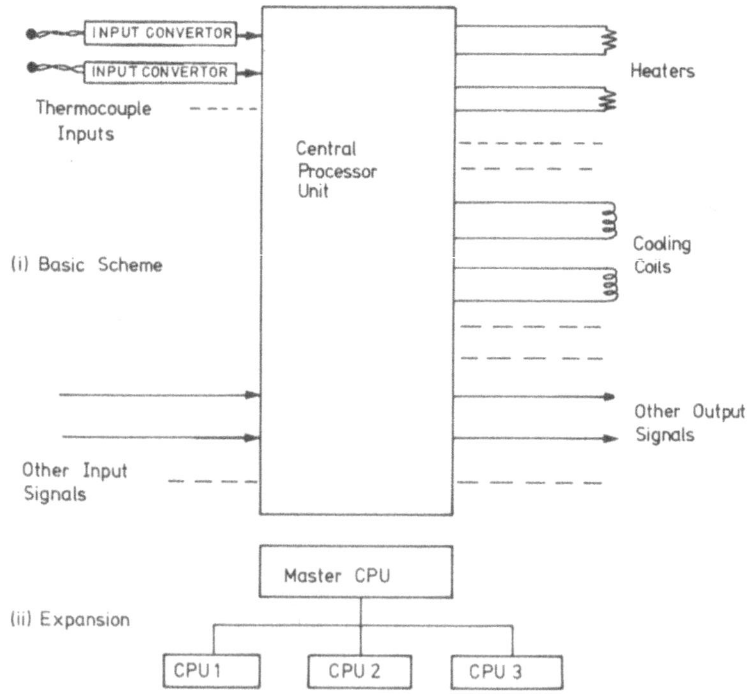

FIG. 21. Direct digital controller for micro-based extrusion control scheme.

functions:

(a) amplification and analogue to digital (A–D) conversion of the transducer output;
(b) linearisation of the resulting digital number to give a real representation, e.g. in units of degrees centigrade;
(c) serial conversion and isolation to give a signal which can go on to a communication bus—in Fig. 21 (i), inside the CPU.

Three-term control action is performed on, say, 16 channels, the exact number being basically determined by timing considerations. Expansion is allowed for by a higher level bus supporting several control modules and controlled by a master controller, as shown schematically in Fig. 21 (ii). This can support a further bus for extra outputs (to cope with the total problem) and communicates with an operator panel.

6.2.1.2. Supervisory scheme. The top level of this scheme, which is illustrated in Fig. 22, is similar to that of Fig. 13 (ii); the basic difference is that now the master controller supervises individual controllers, each of which has its own microprocessor.

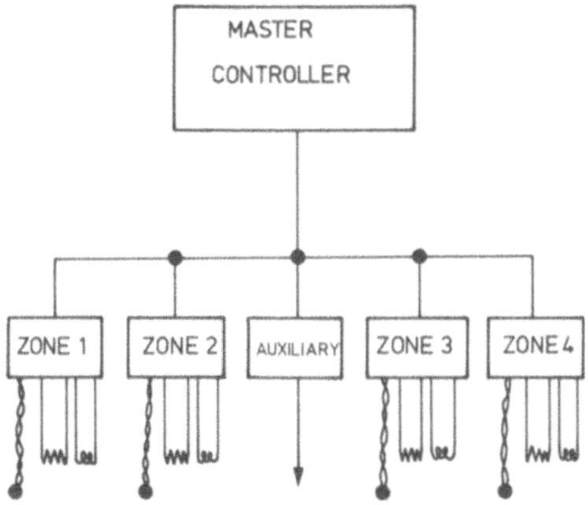

FIG. 22. Supervisory scheme for micro-based extrusion control system.

Comparing the two schemes outlined above, undoubtedly the d.d.c. scheme can be implemented for less cost than the supervisory scheme but it suffers from the following disadvantages.

(i) There is a fair degree of sharing so that failure of a single component can stop part or even all of the temperature loops.

How serious this is depends on the speed at which the fault can be rectified and corrected: it should be noted that there exist many self-diagnostic techniques. A complete back-up scheme such as those used in, say, aircraft control systems could be envisaged but this negates the price advantage.

(ii) Engineers accustomed to the discrete approach will find the d.d.c. approach a very big change and hence commissioning and servicing could be a problem.

(iii) Unless the software for the central controller is written in a highly structured way, maintenance and updating can become a nightmare.

6.2.2. An Industrial Solution

A product, the EM-1, uses an implementation midway between the two extremes giving what is believed to be a very good compromise. Furthermore intensive attention to the operator interface has resulted in a system which gives easy control of the extruder. A block diagram is shown in Fig. 23.

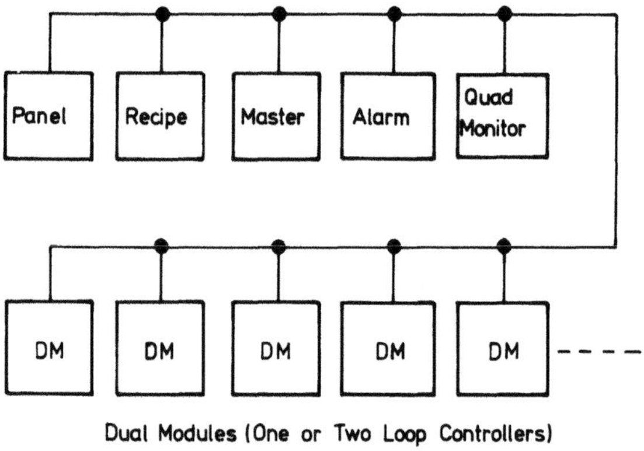

FIG. 23. Block diagram of EM–1.

6.2.2.1. The system. The product consists of a slim, surface mounted Operator Station and a Basic Electronics Package which houses the various plug-in modules that constitute the system. An Extension Electronics Package is available for expansion.

The Operator Station (Fig. 24) has a 32-character display in which each character is a 7×5 dot matrix. This allows full alpha-numeric display and several formats (see below) are possible.

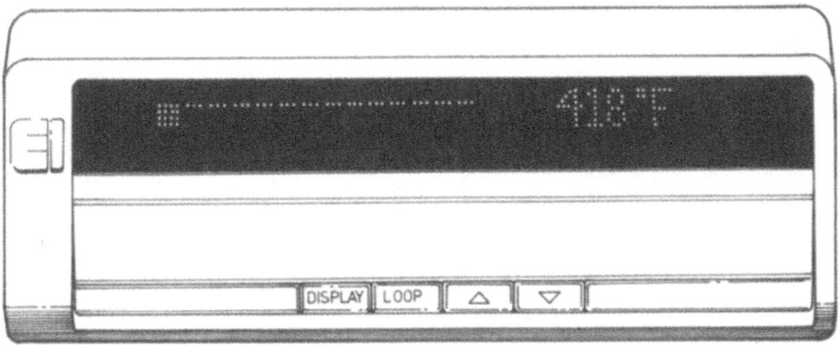

FIG. 24. Operator Station of EM–1.

The number of pushbuttons has been reduced to a minimum and the operator uses only four:

RAISE —to change set-points
LOWER —to change set-points
DISPLAY—to select up to ten lines of process information
LOOP —to select and monitor any control loop

Concealed by a hinged door is a keyswitch and two additional pushbuttons. With the keyswitch in the partial access position, the operator can change set-points between controlled limits and monitor process variables and alarm conditions. With the keyswitch in the full access position the process engineer is capable of changing any control loop parameter by using the PARAMETER pushbutton to select the function and the RAISE and LOWER pushbuttons to change the values. A RECIPE pushbutton allows the engineer to store or recall the parameters associated with a proven product. This feature is very useful, minimising waste due to incorrect procedures and 'twiddling'.

The Basic and Extension Electronics Packages (Fig. 25) are designed for wall or rack mounting in a convenient location. All wiring is done without modules plugged in and each plug-in module is completely protected and is easily inserted or removed. The above units will now be described in more detail.

6.2.2.2. Operator station. The basic principles were described above. The display formats are illustrated in Fig. 26 and are discussed here.

In the *normal mode* (Fig. 26(a)), each temperature zone is represented by a vertical deviation display. If the temperature is high or low in a

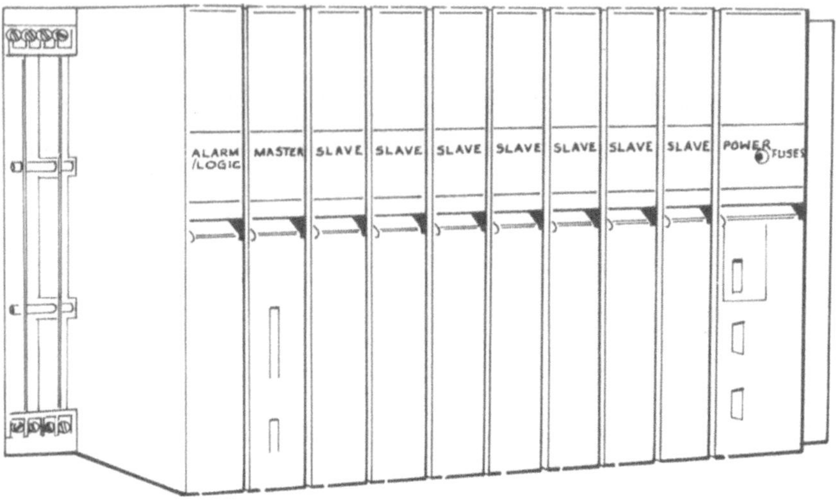

FIG. 25. Electronics Package of EM–1.

given zone, the respective display extends above or below the zero deviation bar. Excessive deviations cause the respective display to flash. This is similar in concept to the controller of Section 5.3.

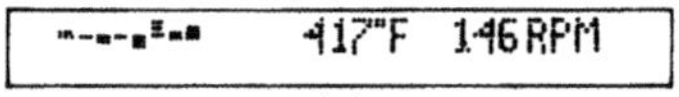

(a)

FIG. 26. Display formats of EM–1. (a) Normal mode; (b) process monitoring; (c) alarm annunciation; (d) control loop monitoring; (e) parameter monitoring; (f) recipe.

For process monitoring, information can be displayed in any convenient form, usually as names and value. Fig. 26(b), for example, shows upstream and downstream melt pressure.

(b)

FIG. 26.—*contd.*

Alarm annunciation is illustrated in Fig. 26(c). Both internal (e.g. loop high alarm, loop low alarm, open thermocouple and loop malfunction) and external (e.g. no cooling water flow, extruder motor overload, barrel

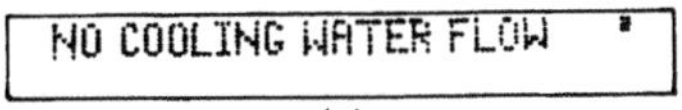

(c)

FIG. 26.—*contd.*

heater failure, sheet tension high, diameter oversize) can be displayed. The operator is notified of alarms by a flashing bar on the extreme right of the display. Depressing the DISPLAY button then shows the number of new alarms and further pressing shows the alarm messages. These alarms are then designated old alarms and can disappear when the alarm condition is rectified.

For *control loop monitoring* (Fig. 26(d)), using the LOOP pushbutton, the measured value, set-point and output level for each loop can be displayed. Set-points can be changed within limits. These limits are preset by a control engineer who has FULL ACCESS. For convenience an acceleration circuit is incorporated so the longer the pushbuttons are depressed, the faster the set-point progresses. This allows one-digit setting whilst still allowing large changes in a short time.

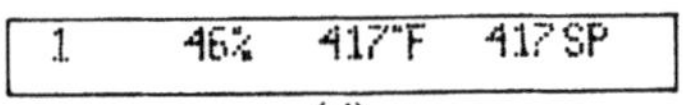

(d)

FIG. 26.—*contd.*

Parameter monitoring can be effected in full access, when all loop parameters can be displayed and changed (Fig. 26(e)) using the RAISE and LOWER buttons. These parameters include proportional band, integral, derivative, high or low alarm, high or low set-point limit, high cutback and low cutback (the latter two are overshoot controls within the digital algorithm). It is useful here to refer back to Fig. 16.

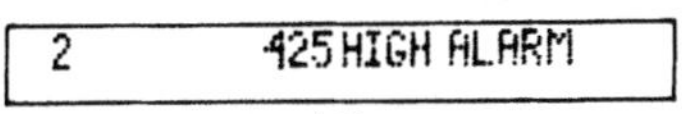

(e)

FIG. 26.—*contd.*

A *recipe* (Fig. 26(f)) is a collection of set-points found to be satisfactory for a particular product: they are assigned the product name. To load a complete recipe requires only pressing the RECIPE button several times till the required product name appears.

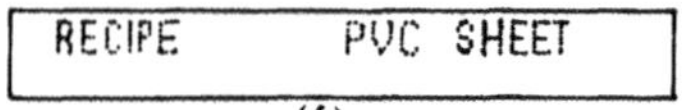

(f)

FIG. 26.—*contd.*

6.2.2.3. Basic Electronics Package. The Basic Electronics Package is essentially a supported motherboard with customer connections into which the modules plug-in. It supplies power, address and an intercommunicating bus to each module. The basic modules are:

Master
Power Supply
Dual Control
Quad Monitor
Alarm Logic
Recipe

A block diagram of the modules is shown in Fig. 23. They are described briefly here.

The *master module* (Fig. 27(a)) drives the Operator Station display and communicates with each module, to assign specific control, monitor or

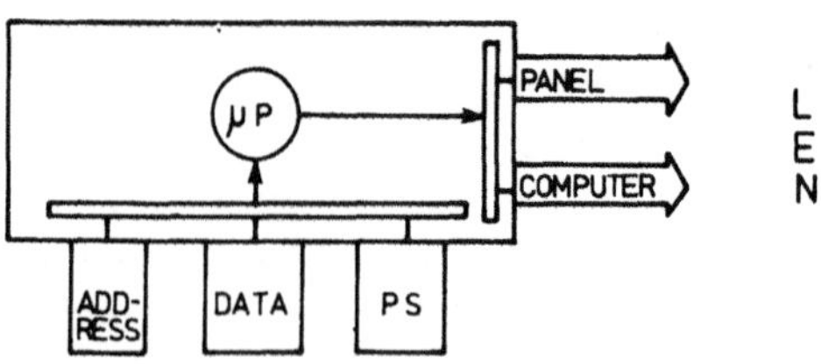

(a)

FIG. 27. Block diagram of modules in EM–1. (a) Master module; (b) power supply; (c) dual control module; (d) quad monitor module; (e) alarm logic module; (f) recipe module.

alarm tasks, via the data bus. It also provides an RS232C or 20mA current loop interface to facilitate a production control scheme such as that shown in Fig. 20. All data are in metalised nitride oxide semiconductor (MNOS), non-volatile memory. The individual control loops are designed to continue to control in the event of a Master failure and an alarm contact is provided for indication.

The *power supply module* (Fig. 27(b)) provides power for all modules plus transducer supplies for pressure transducers.

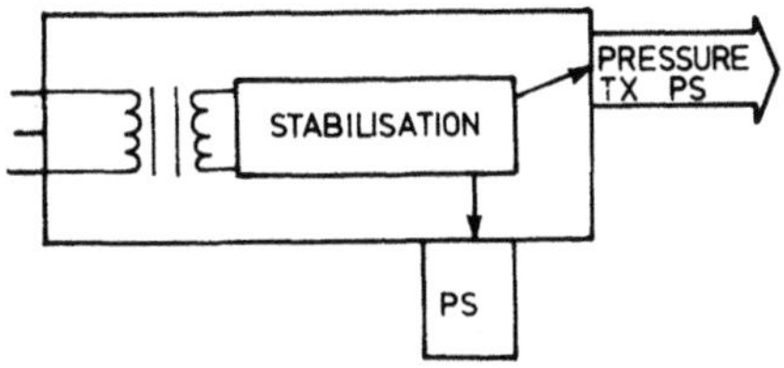

(b)

FIG. 27.—*contd.*

As commissioned by the master module, the *dual control module* (Fig. 27(c)) can perform the following functions:

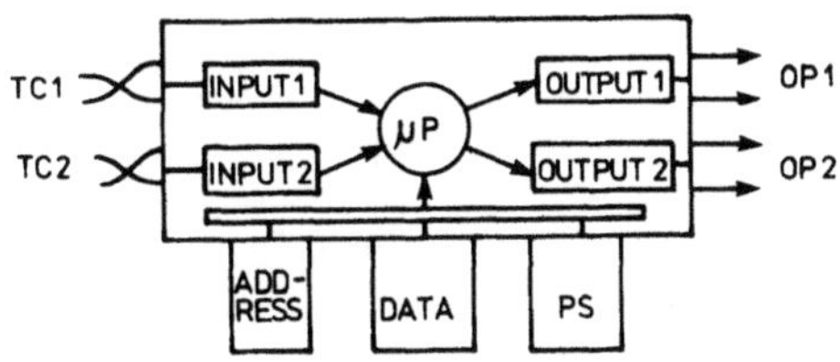

(c)

FIG. 27.—*contd.*

Two heat-only loops	(single couples)
One heat-only loop	(deep and shallow couples)
One heat/cool loop	(single or double couple)

To help provide control integrity, each module has its own microprocessor and an individual LED per channel to indicate the output status. Outputs can be triac, logic or voltage. This versatility should be noted in terms of the discussion above.

The function of the *quad monitor module* (Fig. 27(d)) is to accept up to four external analogue inputs and to communicate these to the master module so they can be displayed in engineering units.

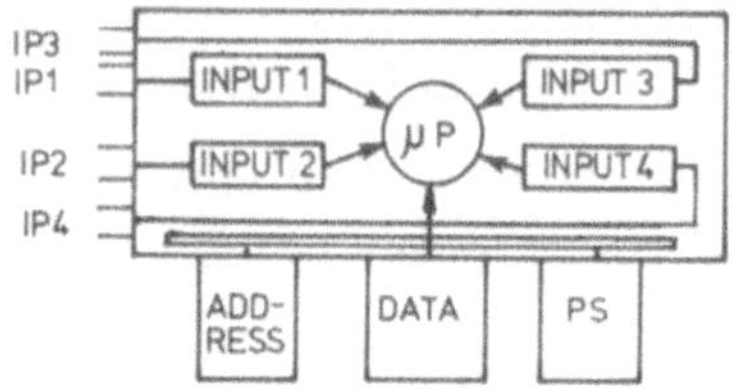

(d)

FIG. 27.—*contd.*

The *alarm logic module* (Fig. 27(e)) has two functions. The first is to provide four relays which can be linked in software to any alarm

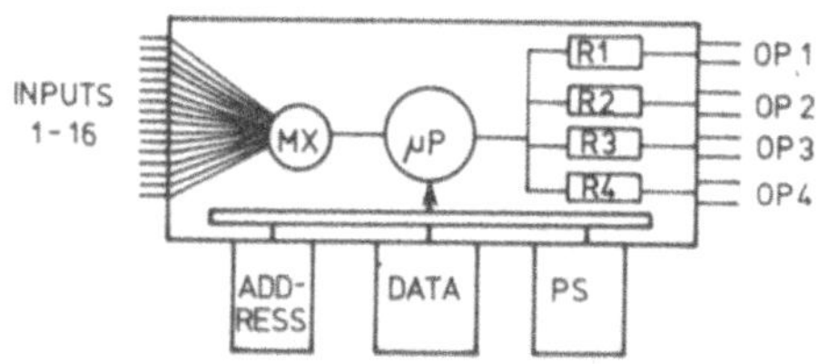

(e)

FIG. 27.—*contd.*

condition arising in the system. These relays can be used for external control functions such as low temperature screw lockout, common high temperature alarm and machine shut-down. The second function of the alarm logic module is to accept as an input up to 16 normally closed external alarm contacts. In the event that one of these contacts opens, this results in alarm display as described above.

The *recipe module* (Fig. 27(f)) is used to store and recall proven process set-up conditions including loop set-points, loop parameters and alarm set-points; MNOS non-volatile storage is used which has a ten-year storage life and is ideally suited to shop floor environments.

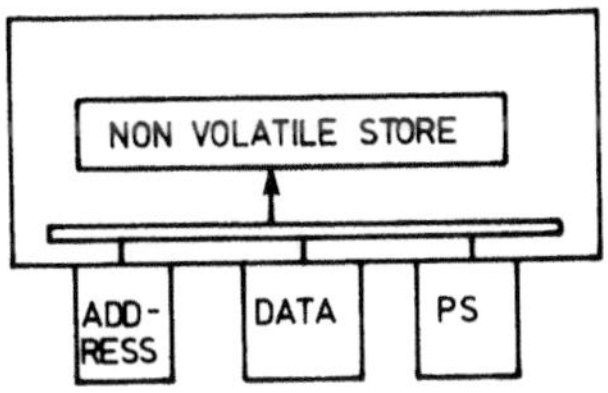

(f)

FIG. 27.—*contd.*

7. CONCLUSION

In this chapter the problems of temperature control in the extrusion process have been considered primarily. In the final section the more global aspect of process control in the extruder factory has been briefly examined. There is no doubt that the movement towards more automation is accelerating as a result of microprocessor technology being applied in all areas—machine control, production control, etc. A key area in this respect is networking of microprocessor-based equipment such as is necessary for total automation. Great strides are being made in this area with the emergence of such systems as Ethernet and we can look forward confidently to advances during the next decade.

Chapter 2

EXTRUSION OF CELLULAR THERMOPLASTICS

K. T. COLLINGTON

FBC Ltd, Hauxton, Cambridge, UK

SUMMARY

Extrudates made from cellular plastics may be produced by using both chemical and physical expansion systems. Both free expansion and controlled expansion techniques are employed to produce pipe, cable, profile, etc. The relevance of each system when based on a particular polymer–product combination is discussed and where appropriate, formulations, processing conditions and corresponding products are given in detail.

1. INTRODUCTION

The level of production of cellular materials using extrusion techniques is small, but a remarkably diverse range of end applications is apparent.

All the major extrusion techniques can be used for the production of cellular extrudates. A range of polymer systems can be processed on conventional or specialist equipment, using either physical or chemical expansion systems in the production of cellular extrudates.

Since the pattern of production and development is diverse, for simplicity, developments are discussed on a process basis, independent of polymer or expansion system used. For reference, the pattern of production and development is shown in Table 1.

TABLE 1

PATTERN OF PRODUCTION AND DEVELOPMENT IN CELLULAR EXTRUSION

Product	Polymer system	Extrusion: Free expansion	Extrusion: Controlled expansion	Expansion system: Chemical system	Expansion system: Physical expansion	Average final density (g/cc)	Comments	Section Ref.
Profile	Poly (vinyl chloride) (rigid)	√	√	√		0·5–1·1	Unplasticised profiles find application in building construction	3.1
	Poly (vinyl chloride) (plasticised)	√	√	√		0·5–1·1	Flexible profiles find application in sealing strips	
	Styrenics (polystyrene, ABS, modified PPO)	√	√	√	√	0·3–0·7[a]	Similar applications to rigid PVC, but the styrenics are more rigid at low densities ABS and modified PPO give better flame retardancy	
	Polyolefins (polyethylene (LDPE), polypropylene)	√	√	√		0·4–0·7	Sealing strips are the main area of end application	
Sheet	Styrenics (polystyrene, ABS)	√	√	√	√	0·01–0·7[a]	Pentane systems widely used in the production of sheet for packaging applications, but high density styrene sheet is also produced, including co-extruded sheet Co-extruded laminates suitable for building panels are produced from ABS	3.2

	Polyolefins (polyethylene (HDPE), polypropylene)	√		√		0·4–0·6	Tapes, binder twine applications, packaging ribbons Vacuum formable packaging materials based on polypropylene Automotive interior trim (talc filled), based on polypropylene	
Pipe and hollow profiles	Poly (vinyl chloride) (rigid)	√	√	√		0·7–1·0	Low pressure pipe systems only Electrical conduit Low density profiles	3.3
	Styrenics (ABS)	√		√		0·7–0·9	Co-extruded ABS pipe available, acceptable to BOCA (USA) regulations	
Cable insulation	Polyolefins (polyethylene (LDPE, HDPE), polypropylene)	√		√	√	0·6–0·8	Single layer insulation based on LDPE in Western Europe Co-extruded 'foam/skin' insulation based on HDPE and polypropylene in N. America CATV cable based on LDPE only Some developments in crosslinked LDPE cable insulation Development in gas injection systems based on LDPE in Japan and N. America	3.3.5

TABLE 1—*contd.*

Product	Polymer system	Extrusion: Free expansion	Extrusion: Controlled expansion	Expansion system: Chemical system	Expansion system: Physical expansion	Average final density (g/cc)	Comments	Section Ref.
Cable insulation *contd.*	Poly (vinyl chloride) (plasticised)	√		√		0·6–0·8	Produced only in the USA, no production apparent in Western Europe	
Tubular film	Polyolefins (LDPE, EVA polymers)	√		√	√	0·6–0·8	Decorative ribbon production Packaging materials (boutique type carrier bags)	3.3.4
	Polystyrene	√			√	0·01–0·7[a]	Pentane or fluorocarbon systems used in the production of polystyrene paper Main application is packaging	
Special processing techniques								
Sheet	Polyolefins (LDPE, polypropylene)	√		√		0·01–0·7[b]	Chemical crosslinked processes operational Irradiation crosslinked processes operational Applications: flotation units; sports goods; automotive insulation	4.1
Hollow profile	Poly (vinyl chloride) NBR blends	√		√		0·3–0·6	Low density pipe insulation, competitive with urethane lagging	4.2

[a] Low density pentane system. [b] LDPE crosslinked.

2. FUNDAMENTALS OF CELLULAR EXTRUSION

The principles of accurate control over melt pressure and melt temperature apply for the production of cellular extrudates independently of the expansion system used. Both physical and chemical expansion systems require processing conditions which maximise solubility of the gaseous expansion system in the polymer melt prior to exit from the die.

2.1. Processing Conditions

Figures 1 and 2 illustrate the optimum conditions for the use of a chemical expansion system in the production of cellular extrudates using conventional single screw extruders (compression ratio between 1·8 and

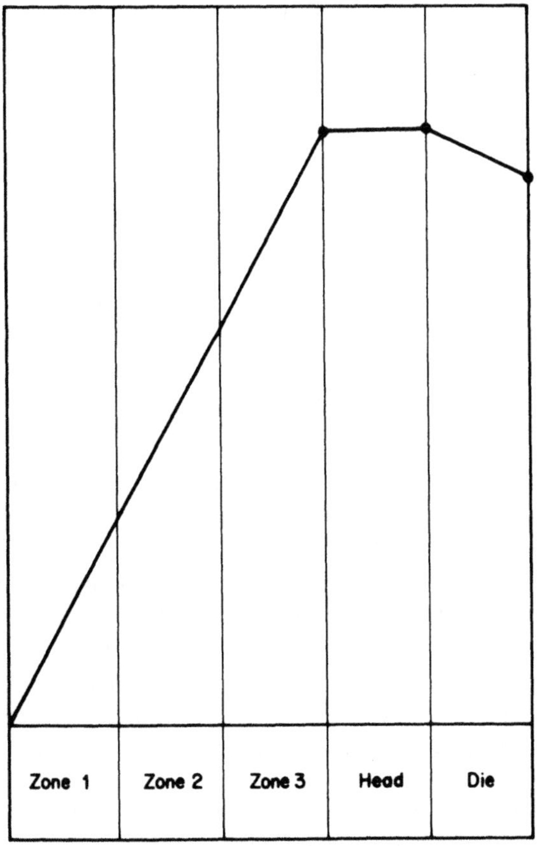

FIG. 1. Melt temperature.

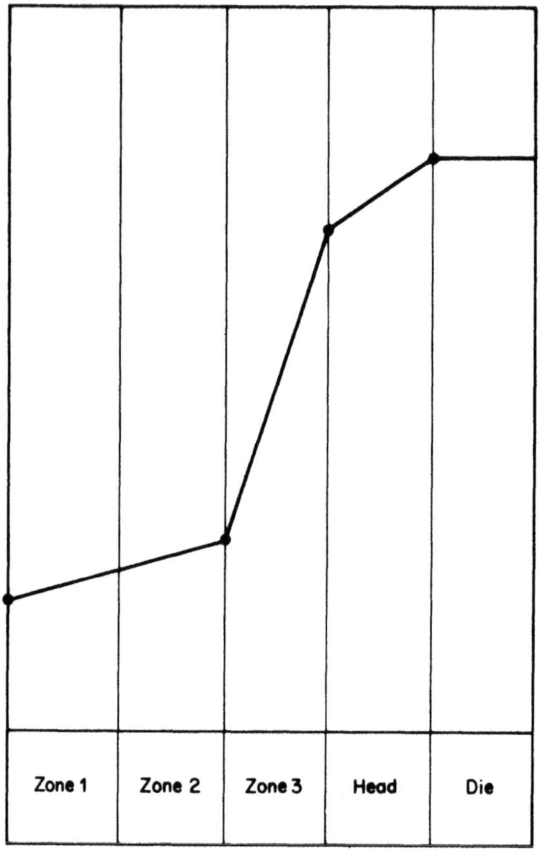

FIG. 2. Melt pressure.

2·5:1). These conditions are also applicable to physical expansion systems based on liquid, or direct gas injection. For processes based on physical expansion systems Fig. 2 refers to the pressure profile beyond the point on the barrel where the expansion system is introduced, assuming a two-stage screw and a homogeneous melt at the point of injection. In both types of expansion system, the object is to maintain melt pressure above the gas solution pressure of the expansion system until the melt passes through the die. Typical melt pressures necessary to maximise solubility of the gas in the polymer melt are 1800–2000 psi (125–140 kg/cm^2 or 12–14 MN/m^2).

Such pressures are within the normal operational pressure range of most extrusion equipment and pose no problems.

Techniques of attaining and maintaining the necessary melt pressure vary, but are essentially functions of screw design, die cross-section and polymer characteristics, i.e. melt flow index and melt temperature. The latter polymer considerations are directly related to the expansion system used. The polymer must be fluid at the temperature at which the gas is introduced, in the case of a physical expansion system, or evolved in the case of a chemical expansion system.

2.2. Expansion Systems

As indicated, both physical and chemical expansion systems are used commercially in the production of cellular extrudates. Theoretically either type of expansion system can be used for the production of a cellular extrudate. However, most development is based on chemical expansion systems with the exception of low density polystyrene packaging materials where a plasticising gaseous system is required to obtain the necessary density reductions and physical properties.

2.2.1. Physical Expansion Systems

The most widely used systems are based on n-pentane, fluorocarbons, or other low boiling point liquids. Considerable development work has been carried out in developing mixtures of fluorocarbons to ensure maximum density reduction by controlling the level of plasticisation in the polystyrene so as to obtain suitable melt elasticity and also to maximise gas retention.

Direct gas injection has also been examined for the production of cellular polyolefin, tubular blown films and slab stock production. It is in this area that initial work on gas mixtures as a means of controlling cell structures and cell size has been carried out.

2.2.2. Chemical Expansion Systems

Both inorganic and organic chemical expansion systems find application in the production of cellular extrudates. The main types of chemical expansion systems are listed in Table 2.

2.3. Nucleation

2.3.1. Importance of Nucleation

Probably the most important factor in the production of cellular extrudates, in particular sheet and tubular blown film, is the degree of control obtainable over cell growth in the polymer melt during expansion. Figures 1 and 2 indicate the melt temperature and pressure

TABLE 2

CHEMICAL EXPANSION SYSTEMS USED IN CELLULAR EXTRUSION

Blowing agent	*Comments*	*Decomposition temp. range (°C)*	*Areas of application*
Sulphonhydrazides *p,p′*-oxybisbenzene-sulphonhydrazide (OBSH) *p,p′*-Thiobisbenzene-sulphonhydrazide (S44)	Low toxicity residues Non-polar residues beneficial in electrical insulation	140–160	Finds application in LDPE CATV mainly Can be used in EVA profiles, also in PVC
Azodicarbonamide and blends	Gives problems of plate-out Can by activated by metal stabilisers	170–210	Finds application in both rigid and unplasticised PVC, where thermal stabilisers activate to reduce the decomposition temperature
(a) Non-plating grades	Eliminate plate-out and corrosion problems Self nucleating	190–220	Find application in most polymer systems independent of process Evolve ammonia, unsuitable for polycarbonate
(b) Activated non-plating grades	As above	160–200	As above, but suitable for high vinyl acetate content polyethylene copolymers
(c) Blowing agent blends	Activation by a secondary blowing agent	160–190	Suitable for polyolefins and thermoplastic rubbers Neoprene LCM systems
Inorganic acid/base systems	Currently pose problems with screw plate-out in polyolefins No problems in PVC	160–200	Main applications are in the polyolefin sheet area and in the 'Celuka' controlled expansion technique
High temperature systems *p*-Toluenesulphonyl-semicarbazide Patented ammonia-free systems	Ammonia-free systems are necessary to eliminate staining problems in ABS and modified PPO	230–235	Profile and pipe extrusions where flame resistance is a requirement

profiles necessary to maximise solution of the gaseous expansion system in the polymer melt prior to exit from the die. However, failure to control the rate at which gas is released from solution in the polymer melt (as melt pressure drops below the gas solution pressure on exit from the die) will result in coarse cell structures and broken surfaces.

Breaking up of the cell structure can be attributed to low melt elasticity and high rate of extension of the polymer melt during cell formation resulting in the rupture of cell walls. For these reasons nucleation is essential in foaming systems based on chemical expansion systems (and physical expansion systems where the gases do not plasticise the polymer melt). In the case of polystyrene, plastification of the melt is obtained by the careful selection of pentane or fluorocarbon mixtures so as to control melt elasticity and gas retention. However, additional nucleation is normally used, i.e. sodium bicarbonate/citric acid, to refine further cell structure and maximise density reduction.

2.3.2. Nucleation Systems

Nucleation systems are based on three types,[1] involving the use of (i) gas mixtures, (ii) fillers and (iii) partially incompatible additives.

Mixtures of gases with differing levels of solubility in the polymer melt can be used to control cell growth. Theoretically gas release is sequential relative to individual gas solution pressures as melt pressure decays. This sequential release reduces the rate of melt extension and cell wall rupture.

Particulate crystalline fillers are used as nucleation systems in polymer melts to control cell formation. On exit from the die the melt temperature of the extrudate drops due to the large increase in volume. However, the volume of the crystalline particles remains constant (with no loss of heat), thus creating localised 'hot spots' where cells can easily form in the areas of high fluidity available.

Partially incompatible additions in the polymer melt, e.g. plasticisers or waxes, can also function as cell nucleators. During foaming, as gas is released from solution in the polymer melt, each cell is encapsulated by the incompatible liquid phase, minimising cell coalescence.

The theory of cell nucleation and cell growth in polymer melts has been discussed in published literature,[2,3] but little practical application is apparent. Practically any combination of the above nucleation systems can be used. Typical systems are outlined in Table 3.

2.4. Extrusion Techniques

Two extrusion techniques are used commercially in the production of

TABLE 3
NUCLEATION SYSTEMS

Process	Expansion system	Polymer	Nucleation system	Comments
Cable insulation	Chemical	Polyolefin	Particulate solid	Source of nucleation is the modified azodicarbonamide
	Physical	Polyolefin	Particulate solid/gas	Chemical blowing agents are used as the nucleation system with the decomposition residues and evolved gas acting as nucleators
Sheet	Physical	Polystyrene	Gas/solid residue/incompatible liquid	Nucleation is attributable to residues and gas evolved from $NaHCO_3$/citric acid reaction. The possibility of excess of citric acid in the liquid phase must also be considered

cellular profiles, pipes and hollow profiles:

(i) 'Free expansion' techniques based on the free expansion of the extrudate beyond the die face.
(ii) 'Controlled expansion' techniques involving mechanical constraint over the rate of expansion of the extrudate.

2.3.1. Free Expansion Techniques

This technique finds application in the production of all types of cellular extrudates and is applicable to all extrusion processes. The free expansion technique (to operate successfully) is dependent on suitable melt elasticity of the polymer system, adequate nucleation and an accurate control over melt temperature and pressure. Melt elasticity of the polymer may be an inherent factor of the polymer system, induced by crosslinking, or resulting from the plasticising of the polymer melt by the expansion system. An example is the production of low density polystyrene sheet using physical expansion systems, i.e. *n*-pentane or fluorocarbons. The pentane or fluorocarbon system increases the melt elasticity of polystyrene, whilst the citric acid/sodium bicarbonate acts as the nucleation system (Table 3), to give flexible, low density foams. Such products cannot be produced with nucleated organic, or inorganic, chemical expansion systems, since the gases normally evolved (N_2, CO_2, CO, NH_3), whilst soluble in the polymer melt, do not plasticise the melt sufficiently to increase melt elasticity.

A disadvantage of 'free expansion techniques' is the problem of maintaining melt pressure above the gas solution pressure, particularly in dies having a cross-sectional area in excess of 5 cm^2, or in complex sections. Development work is apparent in this area to overcome this problem,[4] particularly in polystyrene and PVC expanded by chemical expansion systems.

2.4.2. Controlled Expansion Techniques

Mechanically controlling the rate of expansion of the extrudate (by inducing skin formation and controlling the rate of decay of melt pressure) is the basis of the 'Celuka' process[5] illustrated in Fig. 3.

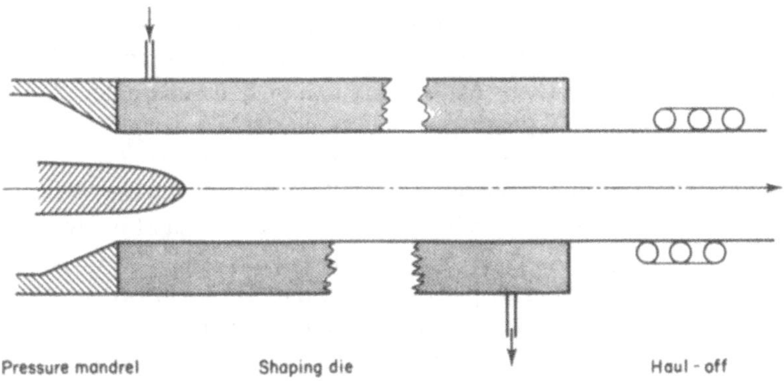

FIG. 3. Controlled expansion system. Expansion of the extrudate is controlled by main die extensions and projecting mandrel. (Reproduced from reference 4.)

Control over the rate of decrease of melt pressure and melt temperature is achieved by the use of an extended mandrel of decreasing diameter. This projects through the main die into a temperature controlled shaping die of increasing cross-section. The system results in the formation of a high density outer sheath for the cellular core which forms as the melt pressure decays—due to the decreasing mandrel diameter and the increasing shaping die cross-section. The main advantage of this expansion system is the ease with which extrudates of large cross-section and/or complex section having hard surfaces, but with low overall densities, can be produced.[6] Typical densities for such extrudates are:

Polystyrene	0·15 g/cm^3 (9 lb/ft^3)
Rigid PVC	0·35 g/cm^3 (22 lb/ft^3)
Polypropylene	0·45 g/cm^3 (28 lb/ft^3)

Only inorganic chemical expansion systems (alone or with the addition of organic chemical expansion systems acting as assumed gas nucleation systems) have, to date, been successfully used in this process. The use of activated grades of azodicarbonamide as used in 'free expansion' systems based on the above polymers has been unsuccessful. The use of physical expansion systems appears feasible, but there is no indication of their use commercially.

2.4.3. Co-extrusion Techniques

The use of co-extrusion systems for the production of cellular cored pipe,[7] cable insulation[8] and sheet is currently an area of considerable interest. This technique can be considered a hybrid system which whilst operating as a 'free expansion' system, exhibits the characteristics of a 'controlled expansion' system.

Control over melt pressure decay is obtained by the choice of suitable melt flow index polymers and the thickness of the outer skin. Practical examples of the technique are in cable insulation where the displacement of LDPE by HDPE and/or polypropylene gave poor cell structures which resulted in poor, or irregular, electrical properties. The use of co-extrusion systems controlled the expansion rate of the cellular core, compensated for the higher fluidity and low melt elasticity of HDPE, reducing cell coalescence to give lower core density, resulting in higher capacitance values.

Co-extruded sheet and simple profiles based on ABS and polystyrene respectively are being produced commercially in North America.

In conclusion, the co-extrusion technique may offer advantages over other extrusion techniques in terms of the cost of die design and raw material savings.

3. EXTRUSION PROCESSES

The diversity of cellular extruded products and the range of polymers used are illustrated in Table 1. Each individual process is discussed below indicating commercial production systems and development in each specific area.

3.1. Cellular Profile Production

Cellular profiles are produced using free expansion techniques based on both chemical and physical expansion systems, and also by the 'Celuka' controlled expansion technique. Unplasticised poly (vinyl chloride) and polystyrene are the most widely used polymers; small quantities of polypropylene homopolymers are also being used.

3.1.1. Poly(Vinyl Chloride)

The production of unplasticised poly(vinyl chloride) profiles probably constitutes the largest area of commercial production of profile.

3.1.1.1. Applications and properties. Main areas of use are in wood replacement applications where the complex sections possible are of interest to the building industry. Such complex sections are used in conditions of high humidity in external applications, or in unit construction, interior panel trim/sealing applications.

The commercial production of rigid PVC profiles is based on both types of extrusion techniques, with the larger cross-section extrudates (exceeding 5 cm^2 die cross-section) being produced by the Celuka controlled extrusion technique. Smaller complex profiles are produced by conventional free expansion techniques using long cooling/vacuum sizing units to maintain dimensional tolerances.

A major area of development in the production of extrudates using free expansion is in techniques to increase surface hardness and reduce density, in line with the economics of wood prices. Average densities produced are 0·8–1·0 g/cm^3 with a starting density 1·4–1·5 g/cm^3. These densities are not competitive with similar profiles produced from polystyrene, which may have a final density of 0·4–0·6 g/cm^3.

Typical physical properties of unplasticised poly(vinyl chloride) extrudates are given in Table 4.

Reductions in density below 0·8 g/cm^3 for rigid PVC result in products with a low degree of rigidity and with soft surfaces which have a tendency to indent easily. Efforts to overcome this problem are based on surface additives (which bloom during processing) and the use of co-extrusion techniques.

Co-extrusion techniques would permit low core densities to be

TABLE 4
TYPICAL PHYSICAL PROPERTIES OF CELLULAR PVC PROFILES

Property	*Test method*	*Density of extrudate* $(g/cm^3)^a$		
		0·67	0·60	0·58
Flex strength (psi)	ASTM D790	3 800	3 150	3 120
Deflection temp. (°C)	ASTM D648			
at 264 psi		57	57	55
at 66 psi		70	70	—
Flammability	ASTM D653	Non-burn	Non-burn	Non-burn

[a] All extrudates of a standard thickness of 0·125 inch.

obtained and could make rigid PVC profiles equivalent in physical properties to both polystyrene profiles (produced using free expansion techniques) and rigid PVC profiles produced by the Celuka process. However, it is doubtful if such production would be economically viable and considerable work is required in this area before commercial production is possible.

3.1.1.2. Formulation. The poor flow properties and low melt elasticity of poly(vinyl chloride) melts are well known and the following factors are being examined to improve production techniques.

Polymer systems are based on low *K* value homopolymers, although both vinyl acetate and speciality ethylene co-polymers are reported to be used in North America. Such polymers obviously impart low physical properties to the final product, but during processing give the rapid fusion characteristics which are necessary to ensure a homogeneous melt at the temperature at which gas is evolved.

Modifiers based on acrylates or ABS are normally used to increase impact strength, but an important secondary function is to increase melt elasticity and refine cell structures. The level of addition is virtually directly proportional to surface and cell quality in the final products, but the economics of formulation cost and end-performance requirements dictate levels of addition.

Lubricants, both external and internal systems based on methacrylates and organic waxes, are normally used to increase gelation of the formulation on the screw and to assist in skin formation, particularly in free expansion techniques.

Expansion systems are based on inorganic systems (sodium bicarbonate) in controlled expansion techniques and on organic systems in free expansion systems and co-extrusion techniques. The selection of the organic expansion system is a function of the polymer grade, but normally activated grades of azodicarbonamide are used. Attention to the selection of thermal stabilisers and other additives must be paid as such additives are capable of interacting with azodicarbonamide to modify the decomposition temperature.

No commercial production based on physical expansion systems is known.

Thermal stabilisers used in free expansion techniques and employing chemical expansion systems based on azodicarbonamide are selected to act both as thermal stabilisers and as activators for azodicarbonamide: they reduce the decomposition temperature to within the processing range of the formulation. Systems used include zinc (Zn), lead (Pb), or cadmium (Cd) inorganic salts or organic complexes,[9] or the more recently developed organotin complexes.[10]

An additional function of the thermal stabilisers is their value as cell stabilisers in both free expansion and controlled expansion extrusion techniques. In the latter extrusion technique their use is essential for the efficient operation of the process.

Fillers and pigments are of value in controlling surface quality and are effective as complementary nucleators. If used with chemical expansion systems the selection of pigments is dependent on the stability of the pigment in the presence of the evolved gases, e.g. carbon dioxide, carbon monoxide, nitrogen and, particularly, ammonia.

A typical formulation, together with basic processing conditions, is given in Table 5.

3.1.1.3. Processing equipment. Little progress is apparent in the development of speciality screws or screw configurations; production is based mainly on single screw extruders regardless of extrusion technique. The major area of development is in die design and calibration systems.

Hollow profile systems are being used where stiff low density sections (density 0·3 g/cm^3) are required. Such dies have land/cross-section ratios of between 5:1 and 10:1 and are designed to minimise pressure decay within the die by the use of restrictor plates.

TABLE 5
TYPICAL FORMULATION AND PROCESSING CONDITIONS FOR PVC PROFILE

Formulation		
Ingredient	*phr*	*Processing conditions*
PVC resin (*K* value 64–67)	100	Melt temperature[c] 160–190°C
Stabiliser/activator[a]	1·5	Melt pressure 1 500 psi
Polyethylene wax	0·2	Vac. calibration unit temperatures to 40°C
Montan wax	0·3	
Calcium stearate	0·5	
Filler	10·0	
Blowing agent[b]	0·6	
Final density 0·8–0·9 g/cm³		

[a] Stabiliser/activator system based on Zn, Pb, or organotin complex.
[b] Azodicarbonamide, or activated blend based on azodicarbonamide.
[c] Melt temperature is controlled by decomposition temperature of the chemical expansion system.

Co-extrusion die design, as illustrated in Fig. 4, is essentially a compromise based on pressure and temperature considerations established for conventional free expansion techniques and the conventional co-extrusion of homogeneous PVC. The main aim is to reduce flow paths (or dwell time of the cellular core within the die block) to a minimum so as to reduce problems of pre-decomposition of the chemical expansion system, and to reduce polymer degradation.

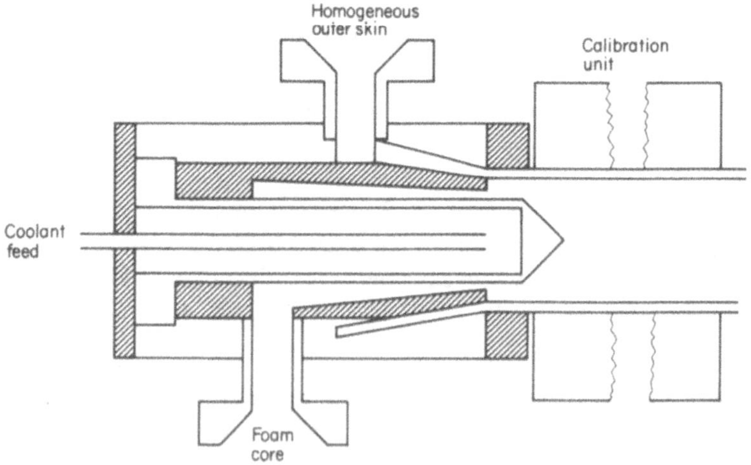

FIG. 4. Co-extrusion die.

Restriction grid dies, as illustrated in Fig. 5, are an interesting development for controlling melt pressure in complex sections. The die cross-section is divided geometrically into small rectangular cross-sections by the use of grid restrictors or numerous torpedo systems within the die lips: such systems maintain uniform melt pressure across the section.[11] A similar system also exists for the production of polystyrene sheet under licence in North America.[12]

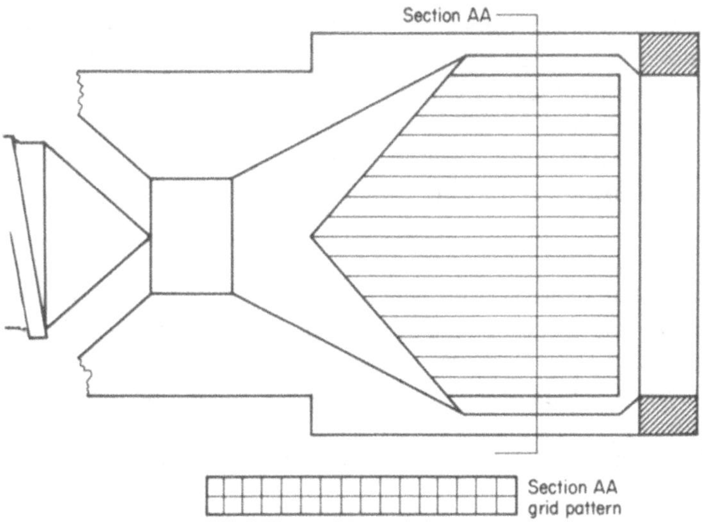

FIG. 5. Restrictor grid die.

Rotating die mandrels are also described in published literature, but no commercial production by this technique is known.[11]

The theory of die design and of calibration units for the production of cellular profiles is still embryonic, but design considerations for homogeneous extrusion appear applicable. The essential factor in free expansion techniques is the formation of a high density outer skin, and calibration units with adequate control over both compression ratios and cooling rates are being designed.

3.1.2. Styrene-based Polymers

Cellular polystyrene profiles are currently produced using both types of extrusion techniques and similar considerations on cross-sectional area (Section 3.1.1) apply in the selection of the extrusion technique. An interesting development was the use of physical expansion techniques utilising gas injection to produce high density polystyrene profiles analogous to the process used for production of polystyrene packing material.[13] Little information is, however, available on the commercial success of this process (initially developed in North America).

Polystyrene offers advantages over unplasticised PVC in both the density reduction possible and surface hardness. These properties are offset by flammability and smoke generation problems which have resulted in developments in both acrylonitrile/butadiene/styrene (ABS) and modified poly(phenylene oxide) (Noryl), both materials capable of satisfying existing legislation.[14]

Little processing information is available in this area, but most die design factors discussed above for unplasticised PVC are applicable and similar extrusion equipment is used in commercial production.

3.1.3. Polyolefins

There is limited production which is based on both controlled and free expansion techniques. Expansion techniques utilise both chemical and physical expansion systems, in the latter case for the production of coarse-celled packaging materials using low boiling point solvents as the gas source.

Little process development is apparent in this area.

3.1.4. Discussion

Savings in raw material costs are probably the motivating reason for continuing development in this area. Increasing legislation, particularly in North America, on burning rates and smoke generation levels, restricting the use of cellular profiles in building applications, could severely restrict further development in this area. Co-extrusion techniques should permit the production of sections with load-bearing properties to extend the use of cellular extrudates further, but such long-term development is based on advances in die design.

3.2. Cellular Sheet Production

Considerable development is apparent in this area based on free expansion techniques using both physical and chemical (organic or inorganic) expansion systems. Development is based almost entirely on polyolefins and styrene polymers with limited developments in the production of plasticised PVC sheet. Development in this latter area is considered transient and related to the current economic situation in certain geographic areas. An example is in South America, where highly plasticised extruded PVC sheet is attempting to compete with conventional PVC coating techniques to produce automotive interior trim, based purely on polymer costs and the availability of paste grade PVC polymers.

3.2.1. Polyolefins

The polyolefins form a significant part of the existing cellular sheet area and could become important in the long term competing with polystyrene in food packaging and with fibre board in automotive interior trim.

3.2.1.1. Current applications. Cellular polypropylene and high density polyethylene sheet are currently used in the production of packaging tapes and ribbons, also in binder twines and weaving ribbons in carpet construction. The essential differences between these products is the degree of draw down used, i.e. draw down ratios greater than 10:1 will give rise to fibrillation, resulting in weaving materials.[15] Polymer selection is based on physical property requirements, aesthetic appearance and handle.

In the case of decorative ribbons physical blends of polypropylene and low density polyethylene are used to obtain the necessary gloss and flexible handle. An additional rheological consideration is the increase in melt elasticity obtained in the polypropylene/low density polyethylene blends, improving cell structures in the sheet prior to orientation. Problems of solids migration, i.e. pigments and particulate cell nucleators, can occur at high draw down ratios and the use of dyes and liquid chemical blowing agent dispersions are of value in obtaining bright uniform colours.

Whilst production of tapes and ribbons can be based on tubular horizontal die systems, commercial production is currently on sheet dies, with average width 40–60 cm.

3.2.1.2. Wide sheet production. Current developments based on polypropylene homopolymers are intended to produce 1·0–2·0 m wide sheet, with an average thickness of 4–6 mm, at a density of 0·45–0·6 g/cm^3. An obvious problem in this area is the control attainable over melt pressure using large width, centred fed sheet dies; these are known to exhibit wide differences in melt pressure between centre line and die extremities. A further problem is nucleation of the sheet to ensure fine uniform cell structure over the conventional parabolic pressure decay curve apparent for centre fed sheet dies. Some success has been achieved in the production of 2 m wide polypropylene sheet using dies designed to minimise melt pressure variations, utilising two-component inorganic expansion systems. The results obtained are interesting, but problems of residual plate out (on both screw and die lips) are apparent and other novel organic expansion systems are now being considered.

3.2.2. Styrene-based Polymers

Whilst the major proportion of polystyrene sheet is based on physical expansion systems (i.e. pentane bead systems, or direct injection of pentane, or fluorocarbon mixtures), developments in high density monolayer and co-extruded sheet (based on both polystyrene and ABS) using chemical expansion systems are apparent.

Applications for these materials are diverse, but the major area of application is in food packaging, where the material savings resulting from co-extrusion are of considerable interest.

3.3. Production of Hollow Profiles and Pipe

The production of electrical conduit and low pressure pipe constitutes the main areas of development, with emphasis on the development of pipe, including co-extruded pipe. Whilst for obvious historical reasons most development is in rigid, conventional single-wall PVC pipe, development is apparent in both co-extruded ABS and PVC pipe. Production of co-extruded ABS pipe is apparent in both North America and Western Europe.

3.3.1. Poly(vinyl Chloride)

Reductions in raw material costs by reducing the density of PVC are obviously attractive, providing physical end-properties are acceptable. Development is apparent in the production of single layer non-pressure pipe based on both controlled and free expansion techniques. Controlled expansion is based on the Celuka process and free expansion is based on patentable specialist die designs.[16]

3.3.1.1. Equipment. In the case of free expansion extrusion, production is on conventional twin-screw equipment and downstream sizing equipment. Melt pressure problems are considerably reduced since uniformity of melt flow is readily achievable providing spiders are of streamlined section. Major problems occur with slight surging and lack of control over skin formation in the bore of the pipe. A result is a degree of corrugation in the bore of the pipe which in service gives problems with retention of solids. This problem has however been resolved by die design technology developed by Armosig, France and subsequently licensed to Robintech Inc., USA.[17]

Co-extrusion based on two- or three-layer systems appears possible and is being evaluated. Obviously no problems of bore corrugation occur when controlled expansion production techniques are used.

3.3.1.2. Formulation. Similar considerations to those outlined in Section 3.1.1.2 apply; lubricants are normally used to delay fusion slightly and to satisfy the processing characteristics of a twin-screw extruder. Substantial savings in unit pipe costs have been reported by the use of formulations containing 11·0 phr of an acrylic modifier; a 30% density reduction could result in a 17·0% reduction in unit pipe costs. Further cost reductions should also be possible by the addition of calcium carbonate fillers, subject to final physical property requirements.[18]

Typical formulations are outlined in Table 6.

TABLE 6
TYPICAL FORMULATION AND PROCESSING CONDITIONS FOR PVC PIPE

Formulation		*Processing conditions*	
Ingredient	*phr*	*(°C)*	
PVC resin (*K* value 67)	100	Barrel zone 1	175
Acrylic modifier	11	Barrel zone 2	198
Stabiliser[a]	0·5	Barrel zone 3	193
Stabiliser/activator[b]	0·5	Barrel zone 4	193
Calcium stearate	0·75	Head	185
Wax	1·0	Adaptor	185
Blowing agent[c]	0·4	Spider	188
Pigmentation	0·5	Mandrel	200
Final density 0·8 g/cm^3			

[a] Conventional thermal stabiliser.
[b] Thermal stabiliser based on Zn, Pb or organotin system acting as an activator for the azodicarbonamide.
[c] Azodicarbonamide only, or an activated blend containing a nucleation system.

3.3.2. *Acrylonitrile/Butadiene/Styrene (ABS)*

Development in cellular ABS pipe is concentrated on co-extruded, cellular cored materials. Details of novel co-extrusion feedblock technology for use in pipe production were released in 1976.[7] Continuing development has resulted in small-scale production with BOCA approval in North America[19] and pilot production in Western Europe.

The co-extrusion technique offers the advantages of improved impact properties and the elimination of problems of bore corrugation. An additional advantage is a saving in raw material costs: a saving of 15–25% at an overall density reduction of 20% has been reported.

3.3.3. Polyolefins

Whilst no work on cellular polyolefin pipe is known, the possibility of co-extruded cellular cored pipe development has been indicated, e.g. for high molecular weight polyolefin systems.

3.3.4. Production of Tubular Blown Film

Cellular tubular blown film with an attractive pearlescent appearance (based on LD polyethylene, or LD polyethylene/ethylene/vinyl acetate copolymers) finds application in flexible packing where high wet strength is of value. Recently a similar film based on high molecular weight polyethylene for replacement of greaseproof paper has been introduced.[20]

Both chemical and physical expansion systems are employed in the production of these materials. Benefits are reported in production using expandable melt concentrates; these give fine cell structures and a more stable bubble form.[21] Development in physical expansion systems is concentrated mainly on mechanical techniques for the introduction of the gaseous expansion system.[22]

Final density of the materials, independent of expansion system, is 0·3–0·4 g/cm^3. Lower densities cannot be attained unless melt elasticity is increased either by inducing low levels of crosslinking, or by including compatible thermoplastic rubbers. The inclusion of the thermoplastic rubbers improves the inherently low tear resistance of the materials, but problems of yellowing of the film have been reported.

A critical control over melt pressure and melt temperature is essential and production development considerations include die design and polymer selection.

Die design developments are based on bottom-fed high pressure systems which offer the possibility of obtaining uniform melt pressure around the periphery of the die.

Polymer selection is critical and low melt flow index polymers (MFI

0·3–2·0) are necessary in order to maximise melt elasticity and 'set up' after cooling. A problem reported in this area during the production of monolayer films is 'die bleeding', resulting in tearing of the film surface. The use of polymers having narrow molecular weight distribution is reported to alleviate the problem, which is attributed to migration of the low molecular weight fractions to the film surface and/or to polymer degradation.

In conclusion, the economics of production of cellular tubular blown film are directly related to the physical property requirements of the end-application. Published work[21] indicates that a decrease in density of 20% will result in a 40% reduction in tear strength, unless EVA or rubber additions are made. Interest in co-extruded materials is apparent and as with flat sheet reductions in raw material costs, novel appearance effects and some insulation characteristics may stimulate further interest in this area.

3.3.5. Cellular Cable Insulation

Developments in this area are in both the micro-coaxial TV cable (CATV) and telephone transmission cable and involve changing patterns in cable construction, polymer systems and expansion techniques.

Both physical and chemical expansion techniques are used in the production of telephone transmission cables, whilst only chemical expansion systems find application in the production of CATV cables.

3.3.5.1. Chemical expansion systems. These are the subject of considerable development work. The systems are based on modified anti-plate out grades of azodicarbonamide and are used[15] specifically in telephone transmission cables. The pattern of development differs depending on the geographic area.

In Western Europe insulation is based on LD polyethylene and this material poses minimal problems in the production of full compounds.[23] Developments in North America involve HD polyethylene or polypropylene homopolymers, both of which may necessitate the use of expandable melt concentrates based on LD polyethylene to introduce the expansion system.[24] The other approach is to utilise speciality polymers which give low head pressures at existing line speeds of 3000–4000 ft/min. Such grades of HD polyethylene are available and permit the use of compounding temperatures below the decomposition temperature of the azodicarbonamide blends during the production of full compounds.[25]

The production of telephone transmission cables using polypropylene is based on 'flake' grades and eliminates compounding problems and the use of melt concentrates. The addition of the azodicarbonamide blends is simply by dry blending; additives are incorporated to reduce segregation during transit or end use.

The changing pattern of polymer usage has resulted in modifications to the particle size of the azodicarbonamide used in the anti-plate out blends. The performance of additives has also been changed so as to satisfy the changes in melt rheology during both compounding and processing.

Developments in CATV cables are less dramatic and (with the exception of micro-CATV production based on azodicarbonamide[26]) the sulphonhydrazide group of expansion systems is used. The main reasons for using the sulphonhydrazides are:

(a) the low decomposition temperature (140–160°C) reduces problems of shrinkage in the thick cellular LD polyethylene insulation layer;
(b) the solid decomposition residues result in minimal loss of electrical properties;
(c) cell quality is good and is attributable to the possible low level of crosslinking of the polymer by the decomposition products.[27] In addition, increased thermal stability has been noted when *p*, *p'*-thiobisbenzene sulphonhydrazide is used, resulting in refined cell quality.

3.3.5.2. Physical expansion systems. Such systems based on gas injection, using chemical blowing agents for nucleation, are the subject of considerable development, mainly in Japan.[28] Distinct advantages in electrical properties are obtained by the use of gas injection or liquid systems, as a result of the elimination of potentially polar residues. However, no information on the economics of the process has yet been published so as to permit comparison with chemical expansion techniques. Development in this area is continuing and in the long term competition with the established chemical expansion techniques is expected.

3.3.5.3. Polymer and process developments. Major differences in the construction of telephone transmission cables are apparent and are responsible for polymer selection. The production of co-extruded 'foam/skin' cable predominates in North America and Japan and is based on HD polyethylene/polypropylene. This contrasts with Western Europe where single-layer, LD polyethylene construction techniques are used. The co-extrusion technique permits the production of lower density cellular insulation (necessary to maintain initial wire diameter when a homogeneous outer skin is added) without loss of abrasion resistance, cracking during braiding operations or compression problems.

Co-extrusion techniques also overcome problems of maintaining cell quality in the less elastic melt conditions existing in HD polyethylene and polypropylene. Controlling the rate of pressure decay, i.e. cell formation, by a viscous outer sheath gives cell structures in HD polyethylene and polypropylene comparable with those obtained in LD polyethylene single-layer foam insulation.[29] Comparison of the two insulation systems indicates that both physical and electric properties are, as expected, controlled by skin thickness and homogeneity.[8]

4. SPECIALIST CELLULAR EXTRUSION PROCESSES

In order to complete this review of cellular extrusion processes it is necessary to include two additional processes in which the extruder is used purely as a method of compounding and shaping a sheet, with expansion as a separate operation. Both processes involve crosslinking, or curing of the polymer system, either immediately before or during the expansion of the polymer melt, depending on the process.

4.1. Polyolefins

The production of cellular crosslinked LD polyethylene based on chemical expansion systems is probably the fastest growth area in extrusion.

4.1.1. Applications of Crosslinked LD Polyethylene

Production of these materials is based on either chemical crosslinking techniques,[30] or the use of ionising radiation to induce crosslinking by chain scission.[31] The technology is of Japanese origin and licensors are operating in both Western Europe and North America.

Applications for these materials exist in the automotive, packaging, pipe insulation, flotation and sports goods areas. Typical physical properties for materials produced using chemical crosslinking systems and irradiation crosslinking techniques are given in Table 7.

TABLE 7

TYPICAL PHYSICAL PROPERTIES OF CROSSLINKED LD POLYETHYLENE FOAMS

Cell structure	95–100% closed cell			
Density (lb/ft^3)	11	6	4	2
Compressive strength at 20°C (psi)				
25%	50	20	10	6
50%	95	40	20	16
75%	250	125	10	50
Compression set[a] (% original thickness)	17	28	85	42
Tensile strength (psi)				
machine direction	244	110	80	42
transverse direction	180	85	68	38
Thermal conductivity (BTU in/h ft^2 °F)	—	0·32	—	0·278
Water absorption (mg/cm^2)	0·6	0·8	0·9	1·0

[a] 22 h compression at 50% deflection, 24 h recovery.

Production involves extrusion of sheet below the activation temperature of the chemical expansion system and subsequent crosslinking and expansion stages.

4.1.2. Production

A typical procedure for the production of crosslinked LD polyethylene foam using chemical crosslinking systems would be as follows:

The formulation would contain the following ingredients:

LD polyethylene
Organic peroxide (dicumyl peroxide)
Expansion system (azodicarbonamide, DNPT (dinitrosopentamethylenetetramine))
Flame retardants
Pigments

Stage 1 Compounding operation (twin-screw extruder, or continuous internal mixer unit). Melt temperature 115°C.

Stage 2 Sheet extrusion. Maximum melt temperature 120°C.

Stage 3 Crosslinking stage (hot air, or infrared combination oven). Maximum air temperature 150°C.

Stage 4 Expansion stage (circulating hot air oven, or combination of infrared heaters and hot air). Air temperature 210–230°C.

Stages 3 and 4 are combined in the chemical crosslinking process with sequential crosslinking and expansion. In irradiation crosslinking no heat is required and this stage comprises exposure of the sheet to ionising irradiation.

4.1.3. Process Developments

Process development is in three distinct areas.

(a) Currently LD polyethylene homopolymers or ethylene/vinyl acetate copolymers are used, but interest is apparent in the use of HD polyethylene and polypropylene copolymers, alone or blended with reactive additive systems.
(b) Polysilane crosslinking systems are being investigated as a possible continuous process,[32] but no commercial production is yet apparent.
(c) Production/material development is occurring in the chemical crosslink systems so as to improve cell quality to match that produced by irradiation techniques. To date, development based on chemical crosslinking techniques has not been successful.

The essential difference between foams produced by chemical crosslinking and irradiation crosslinking is the superior quality of cell structure obtained using irradiation techniques. A limitation of the irradiation process is that on final thickness of the expanded sheet, due to problems in obtaining uniform crosslinking prior to expansion. Excess radiation dosage will cause reversion of the polymer leading to excessive cell wall rupture and high density products.

Whilst chemical crosslinking systems have no limitation on expanded thickness, other than that arising from the thermal instability of the polymer during the expansion stage, cell structures become coarser as expanded thickness increases.

4.2. Nitrile Rubber/Poly(vinyl chloride) (NBR/PVC)

A further large production area is in pipe insulation based on cellular NBR/PVC blends, where expansion is again a separate process either as an in-line operation utilising high frequency/hot air curing systems[33] or as a separate expansion stage using circulating hot air ovens.

Development is aimed at reducing density and improving surface quality. Little information is available on the formulation of such materials but the balance between cure rate and expansion rate is critical if low density products are to be obtained. Reactions between the accelerator systems (used to cure the NBR) and the PVC thermal stabilisers accelerate the decomposition rate of the azodicarbonamide, necessitating a high degree of formulation expertise.[34]

5. CONCLUSIONS

Cellular extrusion offers a production technique for manufacturing a range of extrudates in the form of sheet, profile or hollow tube, of low density and with varying levels of flexibility. Areas of end applications are increasing and it is felt that the use of gas fillers will further extend the level of production for cellular extrudates.

REFERENCES

1. NATURMAN, L., 'How to select blowing agents for thermoplastics', *Plast. Technol.*, October 1969, 41–47.
2. HANSEN, E. H., *Soc. Plast. Eng. J.*, **18**(1), 1962.
3. VILLANIZAR, C. A. and HAN, C. D., *On Bubble Dynamics in Structural Foam Injection Moulding*, Polytechnic Inst., Brooklyn, New York, 1979.
4. COLLINGTON, K. T., 'Cellular PVC' in *Developments in PVC Production and Processing—1*, ed. A. Whelan and J. L. Craft, Chapter 6. Applied Science Publishers, London, 1976.
5. DIRLA, M., *Machinemarkt*, **86**, 1974, 1436–1438.
6. KIESLING, G. C., 'Extruded structural foams', *J. Cell. Plast.*, Nov/Dec 1976, 337–340.
7. NAITOVE, M. H., 'Now co-extrude foam-core ABS pipe', *Plast. Technol.*, February 1973, 50–52.
8. POWERS, J. and DAVIS, L. E., *A Comparison of the Properties of Foam and Foam/skin Insulation in Filled Cables*, Superior Cable Corp., Hickory, S. Carolina, 1978.
9. TU, C. F., 'Dynamic heat stability of expandable semi-rigid PVC, *J. Cell. Plast.*, March/April 1978, 92–96.
10. DVORKIN, R. D., *Decomposition of Azodicarbonamide in the Presence of Organo-tin Stabilisers* (Preliminary investigation), M. & T. Chemicals Inc., Rahway, New Jersey.
11. BARTH, H-J., 'Extrusion dies for rigid PVC foam profiles', *Eur. J. Cell. Plast.*, **2**(3), July 1979, 103–110.
12. ANON., 'Woodlite process, Sekisui patented process', *Plast. Technol.*, October 1974.
13. NAITOVE, M. H., 'Foam profile extrusion excites industry-wide scramble', *Plast. Technol.*, February 1973, 42–44.
14. ANON., *Trans. Cellular Extrusion Committee*, SPE 14 February 1980, BOCA Building Code/1978, Section 920.0.
15. ANON., *The Extrusion of Cellular Thermoplastics*, Fisons Industrial Chemicals, Cambridge, UK.

16. ANON., Technical Literature Ref. N172E, Armocel, Pechiney St Gobain, France.
17. ANON., 'No pipe dream', *Chemical Week*, 23 February 1977, 32.
18. SZAMBORSKI, E. C. and MARCELLI, R. A., 'Designing with plastics—rigid foamed PVC pipe', *Plast. Eng.*, November 1976, 49–51.
19. ANON., 'Foam-core ABS pipe receives BOCA approval', *Plast. Design Processing*, September 1980, 9.
20. British Patent 1525224 to BXL (Flexible Packaging) Division, Darton, England.
21. SHEA, F., 'How to extrude cellular films', *Plast. Technol.*, February 1978, 83–88.
22. Patents to Dutch State Mines, Belgium.
23. NORMANTON, J. K., 'The extrusion of telephone cable insulation using expandable medium density polyethylene compounds', *Trans. ICWS*, December 1974.
24. HODGSON, T. C. and CAREFOOT, D. B., *The Use of Blowing Agent Concentrates to Manufacture Telecom Cables*, Ware Chemicals Ltd, Toronto, Canada.
25. MARSHALL, D. J., 'Methods of inspecting expandable plastics wire insulating compound', *Trans. ANTEC*, New York, May 1980.
26. SUZUKI, F. and MORI, A., *Micro-coaxial Cables Insulating with Highly Expanded Polyethylene by Chemical Blowing Agents*, Sumitomo Electric Industries, Yokohama, Japan, 1968.
27. LASMAN, H. R., *Encyclopaedia of Polymer Science and Technology*, Vol. 2, 532–565, 1976.
28. ORINO, K. and YAMAMOTO, S., 'Gas injection extrusion process for a foamed plastics insulation—supplying a fixed quantity of gas', *Trans. ICWS*, December 1974.
29. MATSUI, M. and MORITA, Y., *The Relation between the Foaming Mechanism and Viscoelastic Properties of HDPE for Expanded Communication Cables*', Dianichi–Nippon Cables Ltd, Amagasaki, Japan, 1968.
30. SANCERNI, J. and LIMONES, A., 'Espumade polietilino reticulado quimicamente', *Plastics Modernos*, No. 221, November 1974.
31. OSAKADA, A. and KOYAMA, M., *Japan Chem. Quarterly*, **5**, 1969, 55.
32. US Patent 4160072 to Sekisui Chemical Co., Japan.
33. SCHINDLER, U., Microwaves in the extruded and rubber goods industry', *Elastomerics*, May 1978, 42–46.
34. MILIGY, A. and KHALIFA, W. M., 'Radiation crosslinking of Nat Rubber, Butyl rubber and CR rubbers in the presence of different additives', *Elastomerics*, December 1979, 28–32.

Chapter 3

BLOWN FILM PRODUCTION

D. J. Dunning

London School of Polymer Technology,
The Polytechnic of North London, UK

SUMMARY

The manufacturing technology for blown plastics film has developed considerably in recent years. Advances have been made in die construction (often with computer aided design), in film cooling methods through internal bubble cooling and dual-lip air-rings, and in process control using microprocessor aided automatic systems. The improvements in machinery have increased film output rate, improved quality and reduced scrap while maintaining the inherent flexibility of the process. Additionally, its scientific principles have been studied, producing a greater knowledge of the rheology of film formation and the effects of processing conditions on film properties. The major tonnage of blown film remains in polyethylene, both high and low density, recently augmented by linear low density. Other blown film materials include PVC, EVA, ionomers and nylons. By blending or using co-extrusion, combinations of materials can confer advantageous properties to the resulting film.

1. FILM EXTRUSION PROCESSES

Since the first blown film plant was patented in 1939 in the United States, the growth of the blown tubular process for manufacturing plastic film has matched the increase in the amount of plastics materials used generally. Above all else the process has been used to manufacture low density polythene (LDPE) film and represents by far the biggest outlet (more than 70%) for LDPE material. LDPE film is used mostly for packaging where, in particular, the changes in the way food is marketed have contributed to an enormous growth in packaging material sales, giving longer shelf life and cleaner products in self-service supermarkets. A sign of the growth rate is the ubiquitous polythene carrier bag, which has only been used widely in the last 15 years.

Basically there are two alternative processes for the manufacture of thermoplastic film, the flat film process and the blown film process.

1.1. Flat Film Process

In this process, melt at a high temperature is extruded through a straight slit die and drawn on to a polished roll. The heat from the film is rapidly removed either by positioning the roll in a water bath (quench process) or by cooling the roll internally (chill–roll process). Adjustments to the flat die lips give overall film thickness tolerance of about ±5% after removal of thicker edges from the film. The flat film process produces a rapidly chilled uniaxially oriented film with outstanding optical clarity and gloss. By installing an in-line reheating and transverse stretching unit (tenter frame), biaxially oriented film can be made. The process is inherently a high investment, high output process. It is used to manufacture polypropylene, polyester and nylon films where control of cooling is necessary in order to obtain the required optical properties.

1.2. Blown Film Process

By contrast the blown film process extrudes a lower temperature melt in the form of a tube which is stretched lengthways and inflated sideways as it is cooled by air directed on to its surface. The tube is flattened into its layflat form and can either be used as a flat tube of film or it can have the edges slit off to produce two single sheets of film. The relatively low melt temperature and the slow cooling employed give rise to a product which is less glossy and clear than the flat film process. Additionally film thickness variations are normally kept to only about ±10%.

The capital investment in a blown film line is much less than for a flat film line (but output is less). Over the years, the investment cost per unit of output has fallen dramatically. Schenkel[1] claims that if present investment costs are assumed to be 100 per unit output, then in 1970 costs were about 250 and in 1955 about 500. The most modern up-to-date plant can produce between 300 and 350 kg LDPE film per hour with a layflat width of about 1 m. Additionally, the ease with which both film width and thickness can be changed gives the process an inherent flexibility. The layflat widths possible from blown film processes range from just over 100 mm to as much as 7 m and thickness from 6 μm up to 250 μm (10 μm $\simeq$ 0·0004 inches, i.e. 0·4 thou or 40 gauge units). The net result of this flexibility coupled with low investment cost has ensured the growth and popularity of the process.

1.3. Film Requirements

It may appear that the technical requirements of a product which is used often for packaging are not particularly stringent. This is not the case. Control of frictional properties, film width and thickness, heat sealability, shrinkage and web-handling characteristics are essential for good performance on automatic package-producing machinery. Equally importantly, there are strong economic arguments for producing the thinnest film possible consistent with satisfactory performance, which means reducing thickness variations within the film to a minimum. Recent increases in plastic material prices have reinforced these arguments.

1.4. Process Development

Much of the development in blown film plant and its operation has centred on the aim of increasing output from such plants whilst maintaining dimensional control, product mechanical and optical properties and additionally diversifying into other plastics which may be more difficult to process. It is the aim of this chapter to describe this progress and how the product properties are controlled.

Apart from its commercial applications, the blown film process has a scientific interest especially in the behaviour of the molten film as it is stretched in two directions while being simultaneously cooled between the die and the freeze line where the film becomes solid. This unique process has been the subject of several experimental and theoretical studies and will undoubtedly be the subject of much more interesting work. These studies will be briefly mentioned, especially where they have most relevance to commercial operations.

2. MARKET STRUCTURE

2.1. Progress of Blown Films

The first successful plastics films were based on cellulose. Regenerated cellulose film, of which small quantities were made in Manchester at the beginning of the century, was produced by a continuous casting process developed by a Swiss chemist, J. E. Brandenberger, in 1911. The process regenerated viscose solution into the cellulose film which was the market leader up until the 1950s. At about this time LDPE became a serious contender. With lower LDPE raw material prices, the development of the tubular blown film process and the introduction of improved film grades, LDPE took over as the leading 'clear' film, especially for packaging. Its good heat-sealability was an advantage but its lower stiffness than cellulose meant new packaging machinery had to be developed.

Stiffer films such as HDPE, PP, polyester (poly (ethylene terephthalate)), nylons and rigid PVC have been introduced successfully together with even softer grades for cling film such as plasticised PVC and ethylene/vinyl acetate copolymers (EVA). The most recent newcomer of significance is likely to be a new type of ethylene copolymer called linear low density polyethylene (LLDPE).

The more detailed processing and product characteristics of these materials are dealt with in Section 6.

2.2. Market Volume

The market for blown films has been dominated to date by LDPE. Total Western European consumption of LDPE plastic film is estimated to be currently in excess of 2,000,000 tonnes per year (see for example refs. 2, 3). In the UK the figure for LDPE film consumption was 330,000 tonnes in 1981. This represents an impressive growth from about 230,000 tonnes in 1975, 130,000 tonnes in 1970 and 280 tonnes in 1951.

A recent breakdown of the UK market is shown in Table 1.

A comparison of major film materials is made in Table 2 which emphasises the dominant LDPE position.

In the last ten years, the market growth of some materials has been more spectacular than that of LDPE in percentage terms but not in absolute tonnage. The market for PP film, especially oriented PP film, has been growing rapidly for the packaging of cigarettes, crisps and other products where a stiff, clear, strong film with good water-vapour barrier properties is required. However, a large part of this demand, especially in

TABLE 1
UK CONSUMPTION OF LDPE FILM IN 1981

Application	*Tonnage*
Heavy duty sacks	35,300
Refuse sacks	25,000
Shrink film	45,000
Stretch film	3,500
Carrier bags	20,000
Agriculture, Building	20,000
Bread bags	15,800
General packaging	165,400

TABLE 2
UK CONSUMPTION OF THERMOPLASTIC FILMS IN 1979

Material	*Estimated consumption (tonnes)*
LDPE	375,000
HDPE	28,000
Plasticised PVC	12,000
Rigid PVC	15,000
PP	30,000
Other	1,000

thicknesses greater than 25 μm, has been met by flat film production.

An area where production of a blown film material has grown rapidly is in that of HDPE films made from low melt flow index (MFI), high molecular weight grades. The amount of this type of film consumed per year in Western Europe was 30,000 tonnes in 1973; it had doubled by 1977 and should have doubled again to about 120,000 tonnes in 1981.[4] Some of this growth has been at the expense of LDPE and some into replacement of traditional paper markets. The higher stiffness and strength of HDPE compared to LDPE means that a thinner film can often be used (with obvious cost advantages).

2.3. Applications of Plastics Films

Applications of films are so widespread that it is only possible to give a very broad survey highlighting some of the more interesting areas.

2.3.1. Applications for Single Layers of Thick Films

Few applications involve completely manual handling of the film and these are often found in the building and agriculture industries. The film is used as an environmental barrier and positioned by hand in such applications as greenhouses, cloches, ground cover (to reduce weed growth and raise soil temperature), frost protection, mulching, waterproof covers, temporary double glazing, demolition site dust containment and damp-proof membranes. The demand is for relatively thick, single thickness films of large widths.

2.3.2. *Shrink Wrapping*

In the field of packaging, which is by far the most important market for film, there has been growth of both LDPE and PVC for overwrapping of boxes, pallets, etc., using both shrink and more recently stretch film methods. In shrink packaging, the orientation left in the film after manufacture is used to give a tight package by heating the film to a temperature at which retraction occurs. For LDPE, film is heated to 110–120°C. For a tight pack, the film should shrink at least by about 60% when unrestrained. Balanced and unbalanced shrinkages are required depending on how the shrink film is used.[5] Shrink film thicknesses vary from 25 to 200 μm though they are commonly in the range 30–38 μm.

Poly (vinyl chloride) film which has been shrunk to fit the product is a very familiar wrapping for greetings cards, etc. Film is heated between 90 and 110°C and, provided its unrestrained shrinkage is in the range 40–50%, a tight clear package ensues. Biaxially oriented plasticised PVC film can be made with the blown film process but commonly monoaxially oriented film is used with orientation being set in the blown film by a separate in-line orientation stage. Smaller items can be packaged on automatic or semi-automatic machinery.[6] An example is an L-sealer and shrink-tunnel. Centre folded PVC film (which may comprise half of a layflat film, slit along its centre line) is opened slightly to allow the product to be inserted between the folded film and then sealed by an L-shaped sealing jaw set at a temperature of about 160°C. This seals the side and end of the package and severs the film in the transverse direction so that both trailing and leading edges are sealed. The enclosed product passes through a shrink tunnel with an air temperature set about 110°C. To allow air to escape during shrinkage, a pinhole can be made in the PVC.

2.3.3. *Stretch Wrapping*

By contrast, the stretch wrap process involves no bulk heating but the wrapping film is stretched, often manually, over the pallet or smaller product and fixed in position generally by heat sealing using a hot plate. The package produced is not so tidy in appearance as a shrink wrap but energy requirements are considerably less. Demand for this form of packaging has grown rapidly over the last five years. Plasticised PVC is particularly suitable due to its high elasticity.

2.3.4. *Sacks*

An extremely interesting area where plastics are in direct competition with traditional paper products is in sacks and bags. The battle in chemical and fertiliser sacks for 25 kg of product has been largely won by low melt flow index LDPE. The sack is produced from the layflat tubular film (typically 200 μm in thickness) by heat sealing in the transverse direction and cutting next to the seal.

2.3.5. *Bags*

Where the waterproof nature of plastics material is an obvious advantage, they again dominate the market. Examples include freezer bags and luncheon bags for maintaining freshness. In the former case, the good low temperature impact properties of EVA have made this material popular though recently there has been a trend to HDPE. Another area of importance is bin-liners which were initially made from LDPE, then HDPE, and now often from mixtures of HDPE and LDPE. 'The mix' comprises about 70% HDPE and 30% LDPE with sometimes a thermoplastic rubber (e.g. EPDM or butadiene–styrene block copolymers) added for toughness. The object of this formulation is the lowest cost product. The stiffness and strength of HDPE enables thinner film to be used provided its toughness and puncture resistance is adequate. Obviously, clarity of film is not a necessary or desirable requirement.

The penetration of plastics into the market for carrier bags and counter bags where the technical superiority of a plastic product is not so clearly demonstrable depends to a great extent on relative prices. In the USA the price structure has meant only a small penetration[7] but in Europe plastics have been more competitive. Low density polyethylene carrier bags are widespread and, despite some initial cases of film of insufficient thickness being used, are now accepted by the public. Manufacture is from gusseted layflat film. Some replacement by HDPE may be expected, though a cautionary tale from Germany tells that economic advantages were somewhat lost when customers took two of the traditionally free carrier bags because experience had led them to believe that the thinner 'plastic' would be inadequate. It will be interesting to see if the LLDPE can make headway in this area.

Counter bags are made from HDPE at very high rates on bag making machinery. This market is very cost sensitive and could well diminish for plastics if their prices rise rapidly. The long-term effect of oil price rises is more favourable to plastics because of the higher energy content of paper but in the short term it is adverse.

2.3.6. *Form, Fill and Seal Machines*

Machines which are fed from reels of plastic film and automatically shape a package, fill it with a dispensed amount of product and then close the package are given the generic name of 'Form, Fill and Seal' machines and represent a large outlet for plastic films. Three major types are vertical, horizontal and thermoforming F.F. and S.[8]

Vertical F.F. and S. machines are the cheapest and widely used, forming a pillow type pack. The film from the reel is fed in an intermittent motion over a forming shoulder to give a tube which is sealed longitudinally and also transversely. Products packaged in this fashion include snack foods, frozen foods, confectionery, groceries, hardware, etc.

In horizontal F.F. and S., the film is V-folded, with the crease at the bottom, and runs continuously through the machine which produces vertical seals, before filling each sachet and sealing the top. These machines are more expensive than vertical pillow pack machines but can operate at speeds up to 500 packs per minute. Powder packaging is a common application.

For thermoforming, the thermoplastic film has to be heated before being drawn into a mould to form a shaped container. The product, such as cheese or cooked meat, is placed in the container which moves automatically to a lidding station where a second plastic film is applied and sealed (possibly under vacuum), before the packs are separated. Composite layered film is often used in these applications.

2.3.7. *Blister and Skin Packing*

Other packaging methods include blister packing and skin packaging. Both require adhesion of the film to the backing card which is used to display product information. In skin packaging, the card is perforated so that the film can be drawn down tightly on to the product. It is one area where ionomer films may be used as their good clarity, excellent draw-down behaviour and resistance to puncture can outweigh their high cost.

3. THE BLOWN FILM MANUFACTURING PROCESS

Before covering later developments, the plant layout and principles and limitations of operation of blown film lines are outlined in this section.

3.1. Extruders

A conventional single-screw extruder supplies a flow of molten polymer to the die which extrudes a thin walled tube, usually vertically upwards.

3.1.1. Screw Design

The *L/D* ratios of the screws are normally in the range 15–25 to 1 and the screws are of conventional three-zone design (feed, compression and metering). For most materials a general purpose screw is used with the lengths of the three zones approximately the same. The depth of the metering zone is dependent on the shear sensitivity of the polymer. LDPE which is not sensitive to shear is processed on a screw with a shallow metering zone giving compression ratios between 3 and 4. However, HDPE which is more shear sensitive is processed with a deeper metering zone and lower screw speeds. This design reduces the likelihood of high molecular weight HDPE being overheated by energy generated during shearing of the high viscosity melt.

For some materials, notably nylons, other screw designs have been recommended. Experience has shown, however, that the general purpose type performs satisfactorily with all materials provided the size of the extruder is sufficient for the required output. Table 3 shows the maximum output that can be expected from screws of different diameters.[9]

TABLE 3
LIKELY MAXIMUM EXTRUDER OUTPUTS

Screw diameter (mm)	*Maximum output (kg/h)*	
	LDPE	*HDPE*
25	45	25
50	160	85
75	340	190

3.1.2. Breaker Plate and Screen Pack

A screen pack and breaker plate assembly is always placed between the extruder and die in order to generate back pressure and hence to improve mixing in the screw. It will also filter out impurities in the melt before they reach the die. Impurities are particularly troublesome in the manufacture of films due to their obvious presence in a thin optically transparent product. In the blown film process they also act as stress concentrators in the blowing stage and may lead to punctures in the bubble with subsequent deflation. A three-or five-screen composite pack with a central screen of 80–120 mesh will filter out most foreign bodies from the melt including some degraded polymer.

The screen pack surprisingly does not eliminate 'gels' of lightly crosslinked or very high molecular weight polymer. These appear in the film as 'fish-eyes' of a size larger than the holes in the wire mesh of the pack. Experiments have shown that at melt temperatures, the gels are sufficiently rubbery to squeeze through the pack in an elongated shape and then return to their equilibrium nearly spherical shape. Because of this ability to cross the screen pack, film grades of polymer are marketed in which the content of gels is low. Generally the level of gels to be expected from a given batch of polymer can only be evaluated by running a sample through the actual production process at normal operating conditions and examining the film produced for fish-eyes. A trial run on a small scale plant can often give different results. These differences between plants indicate that some gels may be broken down in the extrusion process.

The life of a screen pack before it needs replacing depends on the polymer used. For LDPE, with good housekeeping, packs may last for several weeks whereas for a thermally unstable material such as PVC (which also contains additives), screen packs may have to be changed every 12 h. This process often involves stopping the line and separating die from extruder in order to remove the screen pack. Systems are available, however, which do not necessitate extruder shut-down.

3.2. Blown Film Dies

3.2.1. Side Fed Dies

The oldest form of die is of the side fed design and because of its ease of manufacture this type is still in use today. A cross-section through a die of this type is shown in Fig. 1. The melt, after dividing round a central mandrel, flows upwards in an annular channel whose cross-sectional area decreases to the land section. In the land, the cross-section is constant to the annular die exit gap which is usually between 0·5 and 1·5 mm width. The diameter of the exit gap is quoted as a measure of die size and

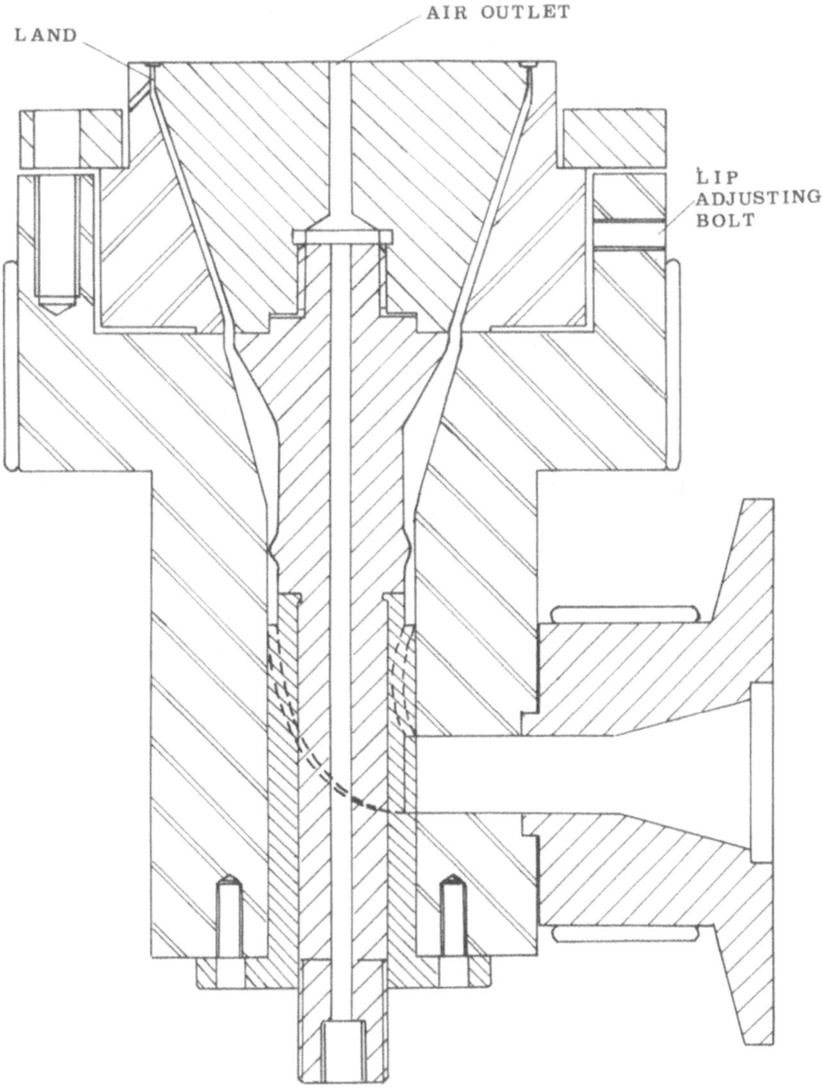

FIG. 1. Side fed die. Based on a design supplied by Betol Machinery Ltd.

ranges from 50 mm for laboratory or small tube dies, through typical commercial sizes in the range 200–300 mm up to speciality dies for heavy duty protection film where the diameter may reach 2 m.

The ring, which forms the outer wall of the land section over the final 30 mm or so, is bolted to the body of the die at a predetermined torque setting so that it can be moved relative to the central mandrel. The torque is selected high enough to prevent leakage between the ring and the die body and low enough to enable the adjusting bolts to shift the ring and so change the gap width at one part of its circumference. The rigid nature of the ring means that the adjustment will produce a corresponding gap reduction on the opposite side of the die.

The design of the side fed die is such that the material flow path on the melt entry (extruder) side of the die is different from the side opposite the extruder. A restriction can be built into the flow channel on the extruder side to attempt to equalise resistance to melt flow for all parts of the channel circumference. However, such a device will not be completely effective over the range of output rates, temperatures and materials which may be used. The asymmetry of the die can lead to asymmetry in the film produced with consequent difficulties in using such film on automatic machinery. Typical problems can be variable shrinkage on heating, a tendency to drift sideways when passing over rollers, and variable thickness giving problems with heat sealing.

3.2.2. Bottom Fed Dies

A more symmetrical design is achieved with the bottom fed die shown in Fig. 2. The melt from the extruder, after turning through 90° in a 'swan neck' type flow channel, hits the underside of the central mandrel or

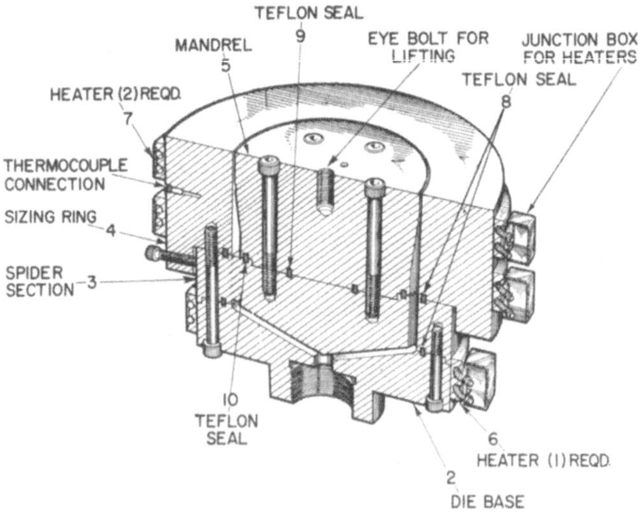

FIG. 2. Bottom fed die. Based on a design supplied by Gloenco Ltd.

torpedo and flows upwards towards the die lips. This design should give an even flow around the circumference of the mandrel. However, to hold the mandrel in place 'spider' arms are required. Normally four are used and they have to withstand a considerable upthrust from the melt. For instance, on a 250 mm die the upthrust for each of the four spider arms would be about 35 tons (approximately 0·35 MN) at a melt pressure of 4,000 psi (28 MN/m^2). The spider arms are thus constructed to withstand these loads. Additionally, they are streamlined to enable the melt to flow round them without producing dead spots in the flow pattern. One of the spider arms is used to take the compressed air supply for bubble inflation to the central mandrel. (However, with this design of die, internal cooling of the film (see section 4.3.2) is not possible.) Despite their streamlined outline, the spider arms break the melt flow and lines of weakness can be produced in the film above each spider. These weld lines become more pronounced as the output rate is increased.[10]

To facilitate melt recombination, the flow channel cross-section is reduced above the spider, thus generating pressure in the melt. However, it is unusual to design for pressures exceeding 35 MN/m^2 (5,000 psi) above the spider arms because of the excessive load which would be placed on the extruder.

Adjustable outer die lips are also fitted to this type of die. The design has proved successful and is still used in plasticised PVC film production where the elimination of dead spots is particularly critical due to the tendency of PVC to degrade.

3.2.3. Temperature Control

Because of their cylindrical shape blown film dies are usually heated by several circular, electrical resistance, band heaters tensioned against the die's external surface. These are controlled in one or two zones preferably with three-term controllers, which sense die temperature from thermocouples set 20–30 mm below the outer surface of the die. The aim in die temperature control is to maintain the die uniformly at a temperature close to the temperature of the melt as it leaves the extruder. For LDPE this temperature is normally about 180°C. The die is not used to heat or cool the molten plastic except occasionally when the outer layers of plastic are heated as they flow through the land. The purpose of this is to improve surface finish by further softening the outer layer of the film before it leaves the die. An additional ring heater and thermocouple are bolted to the top surface of the die for this purpose.

Uniformity of temperature around the die circumference is important in producing bubble stability and for this reason connections to heater bands should be staggered. It is also useful to monitor current flow to each band as the failure of one band can alter the temperature distribution in the die.

3.2.4. Die Exit Gap

For a given film thickness the width of the die gap will influence the total amount of thinning that occurs in the bubble. To prevent frequent blockage and reduce pressure build-up the minimum gap width is usually about 1 mm. For a 25 μm film, this implies thinning by a factor of 40 (not considering swelling of the extrudate which would increase the effective thinning).

Perhaps surprisingly, studies attempting to relate die gap width to film properties have not produced clearly demonstrable effects. Workers at ICI,[5] studying LDPE from dies with gap widths varying between 0·38 mm and 1·02 mm, showed that die gap had only a marginal influence on the shrink properties of the film. HDPE films from dies with gaps between 0·38 and 0·89 mm were tested for tear initiation[11] with the conclusion that no clear trend could be established. Barborek and Williams[12] examined the effect of changing from a 0·64 mm to a 1·02 mm gap on optical and impact properties of LDPE film; the small differences found were probably due to changes in bubble shape between their experiments.

The surface finish of the edge of the die gap drastically affects film appearance with irregularities producing die lines in the film. It is customary to clean the die gap with a soft copper scraper before start-up in order to remove degraded polymer. During die cleaning, the edges of the die gap are especially vulnerable to damage. Small scratches can be honed smooth and larger chips filled by welding before smoothing.

3.3. Cooling Rings

The natural air cooling of the plastic melt above the die is considerably increased by directing a flow of air from the cooling ring on to the outside of the bubble. This is called external bubble cooling.

The air is drawn from the machine surroundings by an electrically powered fan or fans. If a single fan is used the air flows through a distributor where it is split into four to six streams which feed the cooling ring. Alternatively the cooling ring is fitted with four to six separate fans, each with its own air intake. The amount of cooling air is adjusted either by changing fan rotation speed or by use of a restrictor valve in the flowline.

The ring sits on top of the film die with a thermal insulation layer between it and the hot die. Air enters the ring at four to six points on its external circumference and moves inwards over a series of baffles to emerge close to the external film bubble surface between upper and lower lips.

Cooling ring lips are often interchangeable so that the angle at which the cooling air hits the film can be altered. Additionally, the gap between upper and lower lips can be adjusted on some ring designs to give extra flexibility in cooling.

Occasionally, the air is chilled to increase the rate of cooling. In this case it is more efficient to place the chiller unit after the fan as the action of the fan raises the air temperature somewhat.

The design and operation of cooling rings was described by Zielonkowski.[13] It is essential that the cooling air flow is uniform around the circumference of the ring. Non-uniform air flow gives rise to variations in film thickness. This is easily demonstrated by placing an obstruction at one point in the air ring. Above the obstruction, the freeze line rises due to the reduced cooling enabling film at this point to stretch more sideways, thus producing a thinner strip. The strip itself will retract more than average in the transverse direction if heated to produce shrinkage.[14]

The cooling air velocity on leaving the ring should be in a vertical plane through the centre of the die but not necessarily in a horizontal direction.

To achieve suitable air flow, at least four inlets are recommended and these should direct air into the ring either in a radial direction or perpendicular to the plane of the ring. Both methods are found in modern cooling rings as well as tangential entry, which is not recommended.

Even flow over the circumference is obtained by at least three concentric baffles. The gaps between baffles should have a sufficiently large volume to allow even air distribution before air meets the built-in flow resistances between baffles and horizontal surfaces. These resistances should be of a similar magnitude to that at the outlet from the ring.

Raising the ring to leave an open gap between it and the top of the die was found to reduce the cooling efficiency.

Inspection of air rings for damage and build-up of dirt is advisable to ensure even flow. Measurements can be made to check cooling air flow by using a thermal anemometer together with a positioning jib to ensure constant radial and axial position of the measuring probe as it traverses the ring circumference.

3.4. Haul-Off

The haul-off draws the plastic melt from the die and controls the rate of film movement on the plant. It consists of a driven rubber coated roll against which a freely rotating steel or rubber roller is held by a controlled force. The two rolls comprise the nip which holds the two sides of the flattened bubble together to prevent the internal air escaping. To give an even pressure across the nip, pneumatic cylinders should be placed on both sides of the nip. The cylinders may be controlled from near the die to enable the operator to open and close the nip when required. The driven roll is turned by a continuously variable speed motor.

3.5. Bubble Guides and Collapsing Frames

Between cooling ring and top nip, bubble guides may be placed to reduce sideways movement of the bubble. These are often horizontal metal bars (sometimes curved slightly) and are easily movable to accommodate different diameter bubbles. Immediately beneath the nip-rolls, the collapsing frame is fitted to guide the bubble from its cylindrical shape to a layflat sheet at the nips. It consists of two flat surfaces placed on opposite sides of the bubble and directed into just below the gap between top nip-rolls.

3.5.1. Creasing

If the flattening process is not carried out properly the film forms creases as it passes through the top-nip rolls. These creases are difficult if not impossible to remove from the film and apart from their unsightly appearance will give problems in subsequent converting stages, especially heat sealing and printing. An understanding of their origin helps in deciding how to prevent or reduce their occurrence.

3.5.1.1. Origin of creases. If one part of the film is longer and slacker in the machine direction than the rest of the film on the same transverse level, then this part will fold and crease in passing through the nips. Thus it is essential that the die face and top nips are both parallel and that a line through the central die axis passes through the centre of the top nips. These conditions are most easily checked by ensuring nip-rolls and die face are horizontal by means of a spirit-level and that a plumb-line from the centre of the top nips passes through the centre of the die. If the film bubble itself is not symmetrical, with a bulge on one side due to thicker film or lower cooling in that area, then there will be a tendency for this bulge to produce creases.

However, even with a symmetrical bubble and aligned plant, creasing may still occur due to two factors. A geometrical effect may give rise to creases on the edge of the layflat film and a friction effect to creases in the centre of the layflat film. These effects may be understood by consideration of Fig. 3.

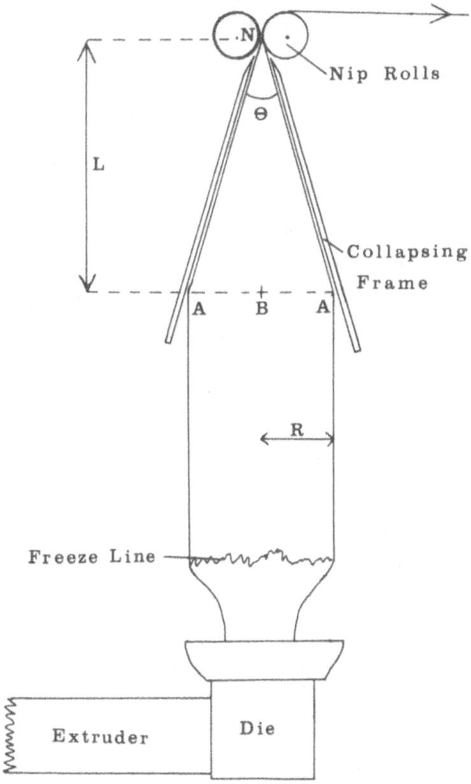

FIG. 3. Geometry of bubble collapsing.

3.5.1.2. Geometric effect. The geometric effect is due to the unequal distance between points on the same horizontal circumference of the bubble and their entry into the nip-rolls. Calculations of the distances to the nip from the base line circumference AA in Fig. 3 show that:

(a) For the film which travels up the centre of the collapsing frame to form the centre of the layflat sheet, the distance AN is given by $(L^2+R^2)^{1/2}$ where L is the vertical distance between the base line and top nip and R is the bubble diameter.

(b) For film which does not contact the collapsing frame but forms the edge of the layflat sheet the distance BN is given by $(L^2+0{\cdot}57R^2)^{1/2}$.

Since the edge distance is less than the centre distance there will be an excess of material at the edge, tending to produce edge creases.

Normally for reasonably flexible films, the centre film stretches somewhat to compensate for the fact that it has to travel further and creases are not produced. However, for stiffer films, e.g. HDPE, this stretching is more difficult, especially when the film is cool. By narrowing the angle, θ, between the frames the *fractional* difference between AN and BN is reduced, thus reducing the necessary percentage elongation of the centre film, and consequently the tendency to produce edge creases. If θ is 40° there is a 4·2% length difference, which is reduced to 1·1% if θ is reduced to 20°. Modern film plant gives facilities for changing θ and for reducing the distance between top nip and die, which will raise the temperature of the film being flattened. Hotter film is obviously more flexible.

3.5.1.3. Friction effect. Creases in the centre of the layflat sheet may arise if the frictional drag between the film and the collapsing frame is large and stretches the film along the line AN compared to other film which travels less distance along the collapsing frame. The frictional drag depends on the coefficient of friction between film and collapsing frame. Plasticised PVC and relatively non-crystalline nylon are two film materials giving high friction. The surface of the collapsing frame is also important. It can be constructed from the following materials, for example, arranged in order of increasing friction: steel, hardboard, linen, velvet. The amount of stretching produced by friction also depends on the stiffness (modulus) and thickness of the film. Problems of centre creases are worst when producing wide, flexible, thin films.

The tendency to produce centre creases will be reduced by increasing the angle, θ, between collapsing frames but this may produce edge creases. Friction will be reduced and the film made stiffer if it is cooler on reaching the collapsing frame. Some manufacturers have used a frame comprising horizontal rollers about 30 mm in diameter. Provided the film does not bulge between rollers this can be successful. For polyolefins (with the exception of thin EVA cling film) centre creasing is not a problem and wooden slats are often used for the frame surface.

3.5.2. *Blocking*

Generally speaking, the layflat film is opened in use and thus it is necessary to prevent the two flattened surfaces sticking together or blocking. Blocking will increase as the film temperature at the nip increases and this factor imposes a lower limit on the height of the top nips. Whilst air from the cooling ring removes most of the film heat, additional cooling before the nip can be obtained by use of water cooled metal surfaces on the collapsing frame. These however will not normally contact the edges of the flattening bubble which will therefore have a different thermal history and possibly subsequently shrink by different amounts. Triangular side plates which may also be cooled can be fitted to contact the sides of the bubble which form the film edges and these sometimes help reduce creasing as well as cooling more uniformly. However, they have to be adjusted at every bubble size change. A method which has had some success is to separate the collapsing frame from the top nip-rolls and pass the film between two cooled metal plates placed in the gap thus formed. This can also help to prevent bulging and creasing of the film between collapsing frame and nip-rolls.

In operation, the bubble flattening process can be treated as an art rather than a science though some understanding of the causes of problems in this area can help with selecting adjustments to the plant. Such adjustments are often necessary when changing from one material to another and sometimes even when changing the dimensions of the film produced.

3.5.3. *Gussets*

For some applications a gusseted film is required to produce, for instance, the base of a bag. Gussets can be introduced in the film by fitting guiding plates on the open sides of the collapsing frame. The plates push the bubble inwards to produce the desired fold as the film passes through the top nips.

3.6. Web Transport

The film web, in its passage from the nip-rolls, which control the web speed, to the winding unit, which controls web tension, is supported by a series of freely rotating flat metal rolls.

3.6.1. *Slitting*

It is possible at this stage to slit the film into several separate webs and/or to trim off the outer edge to separate the two sides of the layflat film. For slitting, the film may be supported by a grooved roll and razor blades pushed through it to fit loosely in the grooves. To produce a smooth edge to the reel of film, slitting should take place as close as possible to where it is wound. Slitting along the edge crease, using knives mounted on a sprung scissor-like frame, will eliminate edge trim but give reels with irregular edges due to fluctuations in film width.

3.6.2. *Web Guiding*

When the film passes over a roll it has a natural tendency to leave the roll in a direction perpendicular to the roll axis. Thus for straight running the roll axis must be kept perpendicular to the required web path. However, due to time-changing variations in thickness and tension across the film (arising often from bubble asymmetry), the film web will move slightly from side to side. This wander can be corrected using web guides which rely for their operation on the film's natural behaviour at a roll.

Initially the position of the edge of the film must be determined using either a mechanical sensor or, preferably, a non-contacting sensor such as an optical or infrared detector. The sensor is placed downstream from the web guide, which is servo-controlled by the sensor signal to maintain the edge in a preselected position. Designs of web guides vary but Fig. 4 shows an example of a swinging roll design.[15] Signals from the edge sensor cause corrective rotation of the frame on which the two rolls are mounted. A frame rotation in the direction X causes the film to move sideways in direction x.

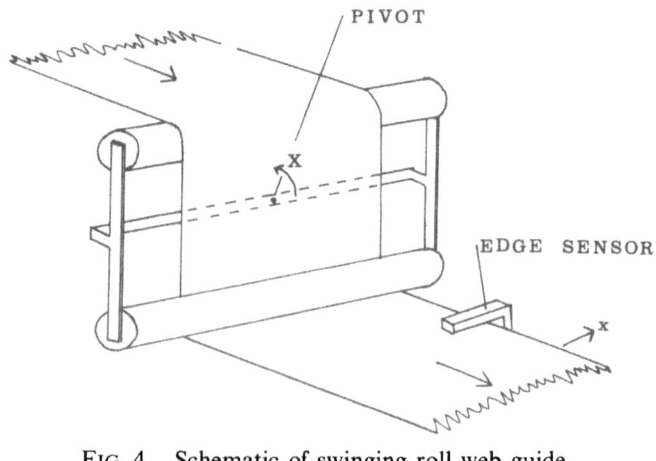

FIG. 4. Schematic of swinging roll web guide.

3.6.3. Spreading

It is important to produce rolls of film which do not contain creases. If creases are present in the film they may be removed by stretching the film slightly in the transverse direction using a spreader roll, which is placed close to the winding unit to reduce risk of creases returning to the film. The distance between the spreader roll and the nearest upstream roll is called the lead-in distance and likewise the distance between it and the following roll, the lead-out distance. For maximum effect the lead-in distance should be greater than the lead-out distance.

The slat roll is used as a spreader roll. Its surface consists of a number of wooden slats lying parallel to the roll axis. Each slat is in two halves which touch at the centre of the roll. As the roll turns the two halves of each slat move apart through half of the rotation and back to their original position through the second part of the rotation. By arranging the film to wrap around that part of the roll where the slats are moving apart a sideways spreading action is applied to the film.

More versatile is the curved or bow roll,[16] shown in Fig. 5. The curve of the flexible roll does not alter its position as the roll rotates about its axis. Thus, since film leaving the roll tries to move in a direction perpendicular to the roll axis, film in the centre of the web travels

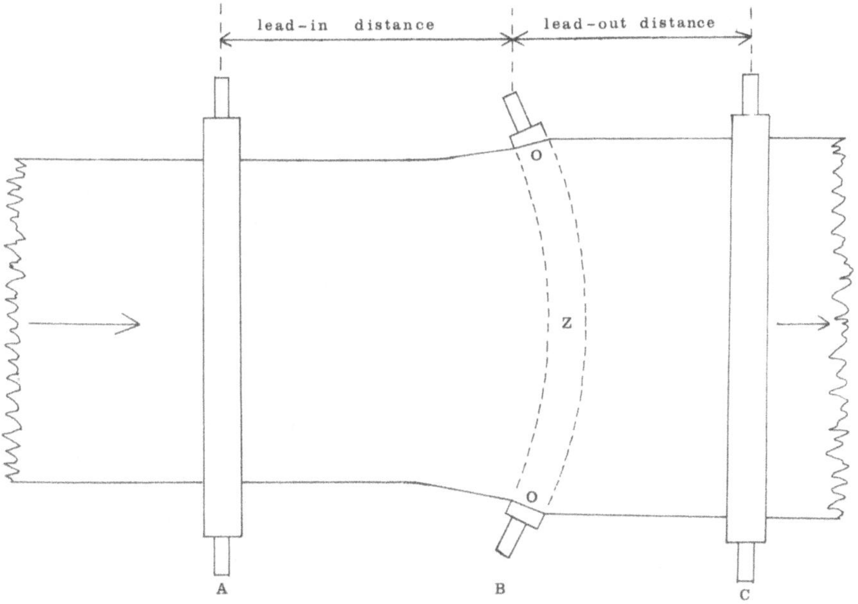

FIG. 5. Bow roll used to apply sideways stretch.

straight whilst film at the edges is pulled outwards. Greater sideways stretching can be achieved by

(i) increasing the curvature of the bow roll, B,
(ii) increasing the distance AB or reducing the distance BC, and
(iii) increasing wrap around, i.e., raising the position of the roll B further above the line joining A and C.

Normally, the plane of maximum curvature of the bow roll in Fig. 5 would be horizontal. However, some films run with either a baggy centre or baggy edges. These faults can be corrected somewhat by tilting the plane of curvature. For a baggy centre the roll is twisted to raise Z into the film; for baggy edges Z is depressed relative to the outer parts O.[17]

3.7. Winding

The film is wound in reel form on a cardboard core (usually 75 mm in diameter, though occasionally larger for 'difficult' films). The core is mounted on a metal shaft using mechanical or air operated end chucks. It is possible to use 'air-shafts' in which a core is loaded and gripped by expanding buttons in the shaft. These shafts are quicker in use but more expensive.

3.7.1. Uneven Film Thickness

Reels of film may show faults due to uneven film thickness and/or to movement of film in the reel after winding. The problem of uneven thickness would occur if a thicker strip of film originating, say, from a thicker section on the bubble circumference, was continually wound in the same position along the reel axis giving rise to a bulge in the film reel at this point. Prevention is by rotating or oscillating part of the line, so that the thick strip of film moves across the reel as winding proceeds and does not build up a bulge at one point on the reel. Present-day plants commonly oscillate the die and air ring through an angle of between 270 and 360°. As the extruder is fixed in position, the die must be fed along its central axis and able to oscillate about the feed pipe. Alternative methods have involved rotating either a horizontally or a vertically mounted extruder and die or oscillating the nip-roll assembly. In the latter case if the wind-up is to remain stationary, the film has to be taken through a concertina-like path in a 'turning-bar' system of rolls. Since thickness variations can originate both from die and cooling ring, the above methods involve relative rotation of both of these items with respect to the winding unit.

3.7.2. *Film Movement in the Reel*

Movement of film within the reel can give rise to buckling near the core and 'telescoping' when there is axial movement of film layers relative to each other. Both effects originate in the inner layers of a reel due to compression from the outer layers. The pressure generated by the compression has recently been measured.[18] Inner layers are pressed closer to the core and their tension reduced by a reduction in their circumference of $2\pi\Delta r$ (Δr is the reduction in layer radius due to compression). If the layer tension becomes negative, the buckling that occurs is seen from the end of the roll as a star-shaped pattern near the core. Lack of tension in the film layers also reduces the force between adjacent layers. Circumferential slippage can occur, often leading to axial movement or telescoping if an axial stress is present. Such axial stresses can arise from thickness variations, lack of parallelism between supporting rollers or tilting of reels during handling and storage.

The amount of compression occurring depends on the inner and outer reel diameter and the tension (i.e. machine direction load) on the film as it is wound on the reel. Harland[19] has analysed reel winding and shown that reel faults should be reduced (a) by increasing the core diameter and (b) by reducing the film tension as the reel diameter increases. The practical effectiveness of (a) was reported by Wanzek.[18] The way in which film tension is controlled is discussed below.

3.7.3. *Winding Systems*

Practically, two distinct methods of winding can be used, surface winding and centre winding.

3.7.3.1. Surface winding. This method is cheaper, enables reels to be changed easily and is the more widely used. The film passes over a fairly large diameter, metal surfaced, driven contact roll against which the growing reel of film is held. The film reel is not driven through its axle but rotates because of its surface contact with the driven roll. The contact roll, provided there is no film slippage on its surface, controls the web tension back to the nip-rolls. If T_m is the torque exerted by the driving motor and R_c the radius of the contact roll, then the web tension, F_w, will be T_m/R_e.

It may be thought, therefore, that surface winding is equivalent to winding at constant tension. There is the possibility, however,[20] that slipping between the outermost film layer being wound and the layer of film beneath may allow the wound tension to be reduced. Slipping will occur if $F_w > \mu L$ where μ is the film-to-film friction coefficient and L is the load between reel and contact roll. In these circumstances wound film tension will equal μL. If the load between reel and contact roll is due to the reel weight, then as the reel weight increases with increasing reel diameter, the wound tension *increases* with increasing reel diameter. Such a reel would be very prone to buckling and telescoping. Contact load is kept constant with reel size by mounting the reel to one side of the contact roll and reducing the slope angle as the reel diameter increases. Additionally, the major load may be applied through bearings with pistons at each end of the reel axle. With these features, the surface winder reels under constant tension.

3.7.3.2. Centre winding. In centre winding, the reel axle is driven and the wound tension controlled by the driving motor. Constant tension winding may be achieved by using a variable-speed motor whose speed is controlled by a signal from either a dancer roll or load cells mounted beneath the bearing-housings of a line roller. Since the driving motor speed decreases as the reel grows, a motor with the correct torque/speed characteristics can also produce constant tension winding if the torque increases in proportion to reel diameter.

Variable winding tension is produced by constant torque winding. Since torque is the product of tension and reel diameter, if it is kept constant the winding tension decreases as the reel diameter increases. Torque is controlled by a clutch or by design of the electric motor and its controls. A reel wound at constant torque will be much less susceptible to buckling and telescoping but its outer layers will tend to be loosely wound. (Harland[19] suggests an even greater reduction in tension with diameter in some extreme cases.) Various 'programmed' winding tensions have been used between constant tension and constant torque to compromise between problems near the core and loose outer layers.

To summarise, centre winding, though more expensive and possibly requiring rotating turret mounting for reel changing, offers the possibility of controlled variable tension winding which may be necessary for some films such as rigid PVC and nylon.[18]

3.8. Process Control

The properties of the film produced are largely determined by the thermoplastic compound used, the thickness of the film, the blow-up ratio, bubble shape and freeze line height. These parameters can all be specified before production commences. In production, the aim is to use the specified material to produce the required film dimensions by using a suitable bubble shape, with output at as high a level as possible. Normally output levels, controlled by screw speed, are selected from previous experience of what is possible. The control of the process is considered in two stages, firstly control action to bring the plant to a state yielding a saleable product and secondly control methods to maintain film quality.

3.8.1. Start-up Procedures

3.8.1.1. Initial steps. In common with all extrusion based processes, before extrusion commences checks are made on extruder and die temperatures, heater band functioning, cooling water flow, air ring operation, nip-roll operation, wind-up operation and raw material supply. It is beneficial to clean the die face and smear with a silicone compound to reduce sticking between the plastic melt and the die and, also, to run a copper scraper inside the die gap. The extruder cooling ring fans and nip-rolls are started at a fairly low speed and after purging for a while the extrudate is attached to a line which is used to lift it through the opened nip-rolls and to the wind-up. Once through the nip-rolls, they can be closed and inflation of the bubble commenced. As the bubble inflates, screw speed, cooling air flow and nip-roll speed are increased to pre-selected values for the product required. The nip-roll speed automatically controls the downstream line speed. Extruder motor current and melt pressure (if a transducer is fitted) are noted and checked against permitted maxima.

3.8.1.2. Setting film width and thickness. The layflat width is measured with a tape and adjusted by the operator either by opening the inflation air valve to increase film width or by judicious slitting of the film to reduce its width by allowing air to escape. The circumferential film thickness distribution is controlled by adjustment of the die gap with the die adjusting bolts. Initially, the operator often checks the freeze line height around the bubble circumference and infers that the film is too thick where the freeze line is high. Adjustments to the die from observation of the freeze line can be quickly made as such observation is extremely rapid. The next stage is to use a mechanical dial gauge fitted with a flat foot to measure the thickness at 16 or 32 points around the film circumference. A sample strip has to be cut from the film, usually next to the winding machinery. If there is no rotation between die and wind-up, adjustments to the die are easily inferred from the measured film thicknesses. Should average thickness be out of specification, adjustments to nip-roll speed can also be made.

3.8.1.3. Other initial checks. At this stage, the operator will also visually check that film appearance is normal, blocking between the two layflat films is not excessive and that freeze line height and bubble shape are as required. After positioning any necessary slitting blades and checking reel widths, the die or nip rotation is started and production of film on fresh cores commences. A yardage meter, often driven from the contact roll, is set to zero and will trigger an alarm when the length of film wound reaches a pre-selected value.

During the first few rotations or oscillations of the plant, a visual check for creasing is made and if present the collapsing frame angles can be adjusted as well as the nip-roll to die distance. The latter adjustment will of course alter the film width and possibly its blocking characteristics.

3.8.1.4. Cooling ring adjustments. Occasionally the film thickness distribution does not appear to respond correctly to the usual die ring adjustments. If the cause of uneven thickness is uneven cooling, then where the cooling is low the freeze line will be high but the film thin because of above-average sideways stretching. This is the opposite of the normal high freeze line/thick film correspondence. Rotation of the cooling ring with respect to the die, followed by a check of film circumferential thickness distribution, should show whether the source of the poor distribution lies in the die or cooling ring. Some cooling rings have top plates which can be tilted to balance the circumferential flow distribution of air. Dirt or the failure of one fan motor may be the cause of poor air distribution.

When the optical, mechanical or shrinkage properties of a film are critical the shape of the bubble and freeze line height may be specified (see section 5). Because of changes in environment, the cooling air temperature will change unless it is controlled. In these circumstances, the operator must also adjust cooling air flow to give the required bubble shape.

3.8.2. Operating Control

During production, reels of film of known length and width are weighed, thus giving a control method for average film thickness. Weighing is one of the more accurate checks on average thickness. Certainly results from dial gauges vary according to gauge and operator so that accuracy is questionable, especially on thin, soft films.

A thickness detector coupled to a chart recorder and positioned after the rotation stage will give a circumferential scan of thickness. If the layflat film is trimmed and rolled into two reels, then it is easy to scan a single thickness. Otherwise the detector is placed close to the edge of the layflat film where the sum of two thickness values at nearly adjacent points on the circumference is registered. The output from the continuous monitoring can be used to give alarm signals if the values fall outside specification. Integration of the signal over one cycle will enable an average thickness to be displayed.

Whilst the film may be fed between a pair of small rollers whose separation is monitored by an electrical displacement transducer, non-contacting thickness gauges are to be preferred for thin films. The most popular is a β-ray gauge, where electrons from a radioactive source on one side of the film surface are recorded after passing through the film. Assuming a constant source flux, the signal from the detector (integrated over a suitable constant time) will be inversely related to the mass of material between source and detector. For a given film material, the signal can be calibrated to give the thickness of a film strip about 5 mm wide. Absorption of electromagnetic radiation may also be used with devices commonly operating at specific polymeric absorption peaks in the infrared region. By placing source and detector on the same side of the film and measuring the intensity of the ray reflected from the far film surface, infrared thickness gauges can be positioned to monitor film thickness before the nip-rolls, giving an earlier warning of thickness changes.

3.8.3. Quality Control

Quality control or assurance tests may also be arranged. Laboratory tests which can be carried out quickly are obviously advantageous and include optical tests such as gloss, haze and fish-eye counts, shrinkage at a selected temperature and time in both machine and transverse directions, MD and TD tensile and tear strengths and elongations at break, tests for block and film-to-film friction coefficients and odour tests. Dart drop impact tests are rather lengthy but a more rapid test is being developed by the Dutch company, DSM (see Section 5). Film gas and water vapour transmission rates are not normally assessed on a routine basis.

3.9. Limitations to Film Thickness and Throughput Rate

It is perhaps useful to mention briefly two aspects of performance which might be expected from the basic blown film line described in this section.

Firstly, the film thickness that may be expected is in the range 10 μm to 250 μm. Below 10 μm handling of the film becomes difficult. Additionally, fluttering in the cooling air stream can give problems. The lowest film thicknesses are easier to reach with stiffer materials—though one application for thin gauges is cling film which requires a soft material. The upper limit is often set by convention and demand and could be exceeded if required. Average film thickness can usually be maintained within $\pm 2\%$ of its target value, but deviations of $\pm 10\%$ from place to place may be expected. This figure of $\pm 10\%$ represents a good control target which should be attainable when operating correctly. Lower variations can be achieved but perhaps not consistently. The percentage variation is expected to increase in thin films. It should be noted that in the case of a 10 μm film a 10% variation is only 1 μm which is probably below the sensitivity of most film thickness gauges!

Secondly, maximum output figures depend even more on the film material and its dimensions. However, for low density polyethylene general purpose film, output is usually limited by film cooling. Typically the maximum figure to be expected is just below 2 kg/h per cm of die circumference (or 10 lb/h per inch of circumference). The numerical value in kg/h is obtained approximately by multiplying the die diameter (in cm) by a factor of 6. Attempts to increase outputs above these figures, when using the basic plant, generally lead to poor gauge control and film creases due to instability of the bubble. Output rates for other materials will obviously differ.

Recent developments discussed in section 4 have enabled the above limits to be exceeded.

4. PROCESS DEVELOPMENTS

4.1. Melt Generation

The extruder must produce a homogeneous melt as cheaply as possible. Developments have centred on increasing output from a given size of extruder, thus reducing capital cost per unit output of melt, and on improving melt homogeneity.

4.1.1. Autothermal Extruders

As extruder screw speed increases, heat generated internally (by shearing of the polymer melt in the extruder) increases compared with heat supplied from the barrel heaters. Adiabatic, or more correctly autothermal, extruders have been introduced. Their screw speed is high (several hundred rpm) and all of the heat necessary to melt the plastic is produced by shearing action in the screw channels. The cost per unit output for such an extruder is low due to high screw speeds, lower-than-normal L/D ratios and simple barrel heating which is required for start-up only. Control of melt output temperature is by a variable restrictor. For example, this may consist of a conical gap between screw tip and barrel with the gap width adjustable by axial screw displacement. Additionally air cooling of the barrel may be utilised. Contrary to makers' claims, lack of control of the melt uniformity has limited the application of autothermal extruders for blown film lines to low cost products where gauge tolerances can be wider.

4.1.2. Screw Design

With conventional extruders, screw design has improved by the incorporation of modifications to improve mixing in the screw and hence melt homogeneity even at high output rates. Studs placed in the flow channel disturb the regular flow path and have been successful with PVC where high shear rates should be avoided. With low density polyethylene, more intensive mixing using high shear sections has proved satisfactory. Their design is such that molten polymer has to pass over the flights in these special sections in order to continue towards the die, thus holding back any unmelted polyethylene. One of these designs, which has been successful in maintaining low melt temperatures at high throughputs, attempts to separate molten from solid plastic by allowing the molten plastic to pass over the flight of a secondary helix which starts at the end of the feed section. The width of the secondary channel increases until it replaces the initial channel by the start of the metering section. Such a design when used with high density polyethylene, for example, gives a melt whose temperature increases only slightly with increasing screw speed.[21]

4.1.3. Grooved Barrels

The flow at the feed end of the extruder has been increased by the use of grooved bushes in the feed zone of the barrel. The grooves run axially, thus restricting the rotation of solid plastic granules at the barrel wall and forcing axial flow towards the die. The depth of the grooves may decrease towards the die and their design depth is critical, balancing increased throughput with the need to avoid overloading a given drive motor. Used in conjuction with subsequent screw mixing sections they give higher outputs of satisfactory melt.

4.1.3. Static Mixers

Melt leaving the extruder may not be uniform, with radial temperature differences of 20°C being possible. It is claimed that such variations can be reduced, say from 20° to less than 5°C, by the use of a static mixer which is inserted between extruder and die. An example is the design manufactured by Sulzer Ltd in which the melt is divided by four consecutive sets of plates mounted in a pipe, edgeways to the flow direction. Each set consists of a number of evenly spaced plates with diagonal corrugations in each plate. Adjacent plates have corrugations running in opposite directions so that the melt is split and recombined in two dimensions as it passes the first set of plates mounted vertically. The second (and fourth) set of plates is mounted horizontally thus achieving a three-dimensional mixing action. The overall length of such a static mixer would be 420 mm for a 90 mm extruder and it is claimed pressure drop across the mixer is low. Mixers of this type have been used on blown film lines to improve melt uniformity and hence film quality.

4.2. Die Construction

Circumferential uniformity of the melt emerging from a blown film die is necessary not only to achieve the required film thickness tolerances but also to obtain high output rates from the plant. Variations around the die circumference of the melt output rate, of the melt temperature or of the shear history inside the die lead to asymmetry in the bubble. Consequent variations in film tension, thickness and shrinkability give problems in collapsing the bubble and in film handling and conversion.

Die designs have thus developed from the side and initial bottom fed designs. A star fed design led to the spiral mandrel type which is now the most successful type of die.

4.2.1. Star Fed Die

In this design, the central mandrel is fixed to the die body at its base, thus obviating the need for spider arms. As shown in Fig. 6 the melt is piped under the die and turned through 90° to move upwards along the central axis of the die. The flow channel is then split into about six

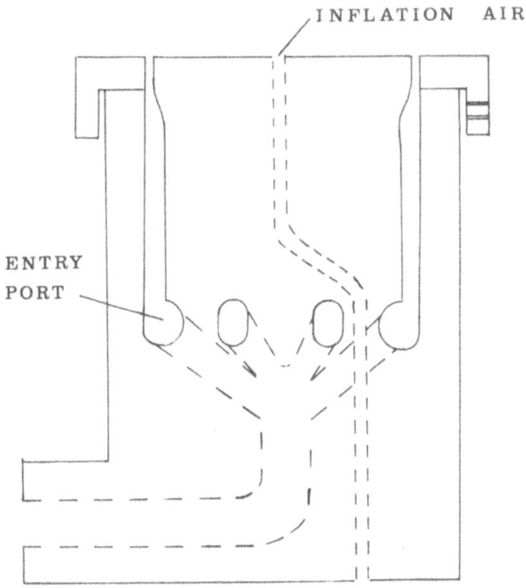

FIG. 6. Star fed die.

separate arms which enter the base of the plenum between mandrel and die body. Entry is through ports spaced evenly around the circumference of the mandrel. Melt then flows up through the plenum whose annular cross-sectional area decreases towards the die lips.

Such a die, while achieving gross circumferential symmetry, can give problems with surging at points above the entry ports and incomplete welding of melt flows from adjacent ports.

4.2.2. Spiral Mandrel Die

4.2.2.1. Mode of action. The spiral mandrel is a development from the star design. Melt from the separate entry ports is spread circumferentially and blended together by means of spiral flow channels cut in the central mandrel. As the flow channel spirals up the mandrel from the entry port, its depth decreases, as shown in Fig. 7. Additionally, the gap between mandrel and die body increases with a consequent smooth transition from flow in the channel to upward flow over the lands between the channels. In this way melt from each entry port is spread around the die circumference and blended with melt from adjacent ports.

The exit from the die is fitted with a standard adjustable ring in order

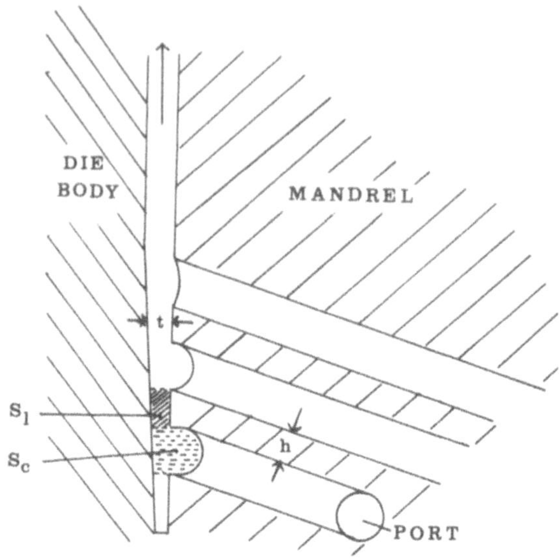

FIG. 7. Base of spiral mandrel. S_l=land flow space; S_c=channel flow space.

to vary the die gap width when necessary. Such adjustments are carried out by bolts which give rather a coarse and sometimes jerky shift to the gap. Considering that output from a slit gap over which there is a constant pressure drop is proportional to the third or higher powers of the gap width, this crude adjusting method is not altogether satisfactory. Some development work has been carried out on alternative means of adjusting the ring such as thermal expansion of the bolts. However, this has not yet been brought to a commercially successful conclusion.

Nevertheless, the spiral mandrel is now the most common and successful die for blown polyolefin films. Its design also accommodates internal bubble cooling air flow channels (see Section 4.3).

4.2.2.2. Design. The designer of spiral mandrel dies has a wide choice of variables, principally the rate at which the spiral climbs the mandrel, the rate at which the depth of the spiral channel decreases, the variable size of the gap between mandrel and die body and the resistance to flow between the top of the spiral channels and the die lips. To assist in design, it is possible to analyse the flow of polymer through spiral mandrel type dies and from such an analysis to predict the behaviour of a given die design. The effects of changes in design can then be predicted from such calculations.

The criteria for evaluating a given design include the following items:

(a) Uniformity of output around the die circumference. A narrow gap between mandrel and body can cause the bulk of the plastic melt to flow around the spiral groove until near its termination point and then give rise to a thick film spot above each termination point.
(b) Mixing of melt from adjacent entry ports. If the gap between mandrel and body is too large, there will be little flow along the spiral grooves, little mixing of melt from different entry ports and possibly consequent weak lines in the film at the boundaries between adjacent melt streams.
(c) Correct overall flow resistance to the design melt output rate. This will ensure sufficient back pressure on the extruder without overloading.
(d) The melt shear and shear rate to be largely independent of flow path. This is a necessary condition for melt temperature uniformity due to possible shear-heating of the melt.
(e) Volume of melt to be as small as possible. Together with (d) this assists in rapid purging of the die.

Die performance depends on the interaction of polymer melt and die geometry. Therefore, before summarising the method of analysis, it is necessary to consider how a particular plastic material can be characterised for the purpose of die design.

4.2.2.3. Melt characterisation. Published attempts[22,23] to analyse the flow through a spiral mandrel die have assumed that the flow behaviour of the plastic melt can be described by a Power Law equation. These equations assume that the shear stress, τ, is related to the shear rate, $\dot{\gamma}$, by eqn. 1.

$$\tau = K\dot{\gamma}^n \qquad (1)$$

K and n are constants over the range of shear rates under consideration.

A most suitable method of obtaining polymer flow properties is by capillary rheometry. The volume output rate, Q, from a capillary die of radius R, and length L, is obtained as a function of the pressure drop, ΔP, across the capillary. It can be shown (e.g. ref. 24) that if a plot is made of $\log\left(\frac{R\Delta P}{2L}\right)$ as ordinate against $\log\left(\frac{4Q}{\pi R^3}\right)$ and a straight line fitted to the data, then, for polymers which obey eqn. 1, the slope of the line gives the flow index, n, and its intercept a consistency constant, K', related to K by eqn. 2.

$$K' = K\left(\frac{3n+1}{4n}\right)^n \tag{2}$$

The constants n and K' may be used to characterise a specific grade of polymer at a given temperature. Values of n vary from 0·2 to about 0·7. In film grades, n is generally lower for polypropylene than for LDPE, whose n value is lower than that for HDPE.

4.2.2.4. Flow analysis. It is assumed that the space between mandrel and die body may be divided into two regions as shown in Fig. 7. In the channel flow space, eqn. 3 is assumed to hold.

$$\frac{\Delta P_c}{l} = \frac{2K'}{r}\left(\frac{4Q_c}{\pi r^3}\right)^n \tag{3}$$

where ΔP_c is the pressure drop along a length, l, of the spiral channel, Q_c is the flow rate along the channel, and r is the equivalent radius of the channel.

For flow over the lands, eqn. 4 is assumed.

$$\frac{\Delta P_l}{h} = \frac{2K'}{t}\left(\frac{6Q_l}{wt^2}\right)^n \tag{4}$$

The height of the land is h (see Fig. 7) and t is the gap between land and die body; ΔP_l is the pressure drop across the land when the flow rate is Q_l for a distance, w, around the mandrel circumference.

To understand the basis of the calculations, which are normally carried out by computer, the outer surface of the mandrel is 'unrolled' as in Fig. 8, which shows a design with four entry ports. This shows that since points B and C are equidistant from the entry port along identical spiral channels they should be at the same melt pressure. Hence the channel pressure drop between A and C should equal the land pressure drop between A and B.

The computer is firstly given an assumed pressure profile along the channel, e.g. a linear pressure decrease between entry port and termination point. The channel is divided into a number of sections and the computer calculates flow rates over the land for each section using eqn. 4. Flow in the channel is calculated from the flow out of the entry port by

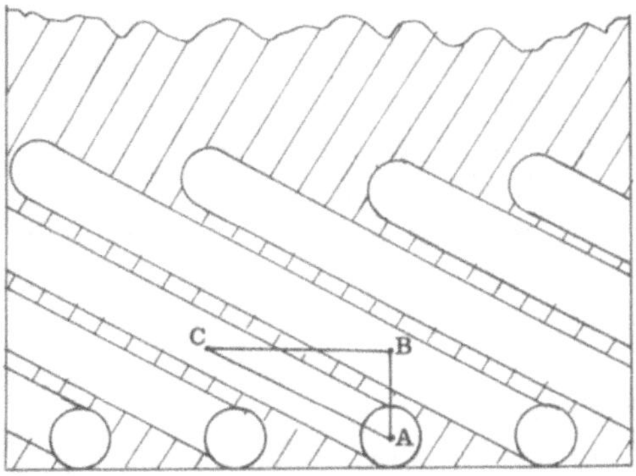

FIG. 8. 'Unwound' four-port spiral mandrel.

subtracting land flows from each section between A and C. Beyond C the channel flow is reduced by upward land flow but reinforced by flow from beneath. The resultant channel flow is used (eqn. 3) to calculate a new assumed channel pressure profile to replace the initial assumption. The whole process is repeated until iteration gives consistent results.

The computer will then tabulate net upward flow rates at different points on the mandrel circumference, the channel and land flow rates within the die, the shear rates in the channel and the land, and the pressure profile in the die.

4.2.2.5. Conclusions from analysis. Kurzbuch[23] tabulates results from a 'typical' eight-port die in which each channel spirals around the mandrel about twice (720°) before terminating. The results show that, in this die, the majority of channel flow should convert to land flow between 60 and 180° from the channel entry port. This could lead to polymer 'hang-up' in the remainder of the channel. A die which gave a more even transition from channel to land flow was designed and built with claimed excellent performance in purging and thickness distribution.

Procter's results[22] indicate improved designs obtained by using a relatively deep initial channel, a relatively steep spiral climb and little or no increase in the mandrel-to-body gap width beyond the initial widening. His analysis indicated that polymer grades with low values of flow index n should give more uneven output from a given die, but this was not clearly demonstrated by experiment.

The success of the above approach is borne out by the fact that many companies who manufacture plant for blown film production design their spiral mandrel dies using computer analysis of the flow.

A recent paper by Cheng[25] demonstrates how later die designs have been influenced by these calculations. In the dies shown, there are between eight and twelve entry ports leading to spiral channels which climb steeply and terminate after 180–360° of turn around the mandrel. To achieve uniform flow from the die lips, Cheng suggests designing the spiral channel and mandrel body clearance in such a way that the leakage rate over the lands from the channel is constant along the length of the spiral channel.

4.3. Cooling Systems

An increase in film cooling rate, without an adverse effect on thickness tolerances or bubble stability, should allow an increase in output from a given blown film line. Consequently studies have been carried out on film cooling resulting in some improvements when using the standard air cooling from outside the film. This is described in section 4.3.1. A development of greater significance involving internal cooling of the bubble with its attendant modification to the die is covered in the following section.

4.3.1. External Cooling

4.3.1.1. Air flow pattern. Experimental studies of air flow rates and temperatures and their effect on bubble temperatures under a range of conditions have been undertaken in Germany.[14,26–28] It is found that air leaving the lips of the cooling ring at a velocity of 10–30 m/s rapidly achieves a characteristic flow pattern moving in the same direction as the film (whose velocity will initially be very much lower, e.g. 0·01–0·1 m/s at the die exit). The air closest to the film develops a laminar flow layer which increases in thickness as the distance from the die increases. The conduction of heat from the film into this layer increases with its velocity and falls as the temperature of the air increases and the film temperature decreases. Outside the laminar flowing layer there is a turbulent mixing of cooling air with the atmosphere. Together with the film friction this reduces the velocity of the laminar flow layer as it moves away from the die.

4.3.1.2. Cooling rate. Infrared thermometric measurements show film temperature falls nearly linearly between a point just above the die and the freeze line. Raising the volume of cooling air flow and its velocity increases film cooling rate. However, deformation of the bubble and 'fluttering' limit the extent to which this method can be used to increase maximum film throughput.

It was found that the angle at which the cooling air was directed from the air ring only marginally changed the cooling rate. Slightly faster cooling was apparent when using a low cooling angle (i.e. close to normal to the film and parallel to the top of the die). However, this more rapid cooling is offset, especially in low viscosity melts, by a reduction in the maximum air velocity possible before bubble instabilities occur. The tendency has been therefore to increase the angle so that the air flow from the cooling ring lips is parallel to the film surface. At the same time the upper lip can be extended vertically for a short distance producing a narrow annular air stream between film and lip. At the top of the upper lip an eddy is formed which disturbs the laminar flow and is an important factor in raising cooling rates. Beyond the cooling ring the air stream will support the bubble to a certain extent.

Later disturbances in the flowing air can be introduced. A wire ring supporting many small PTFE rollers can be positioned next to the film surface. This will force the air to flow out round the ring and increase cooling by disturbing its flow. However the positioning of such a device is difficult, especially if film diameters are not constant. An iris diaphragm, placed round the film bubble above the cooling ring, with the iris diameter somewhat greater than the bubble diameter, will introduce an external disturbance into the laminar flowing air layer. This is reported to allow an increase in output of up to 10% but can cause a deterioration in uniformity of film thickness.

4.3.1.3. Dual lip rings. Greater improvements can be made by use of a second annular air stream impinging on the film above the first. Figure 9 shows a cross-section of such a cooling ring manufactured by Overbeck and marketed in the UK by Kirwan Machinery. The cooling air after passing over the normal equalising baffles is divided into lower and

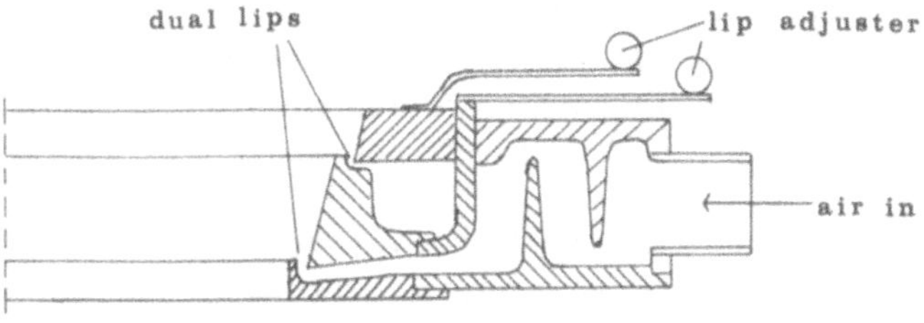

FIG. 9. Cross-section through dual lip cooling ring. Based on Overbeck design.

upper streams which leave the rings vertically one above the other. The diagram also shows the high upper lip which also contributes to the cooling function. United States experience of such dual-lip rings shows[29] that film output can be increased by as much as 30% compared with single-lip cooling.

4.3.2. Internal Cooling

For a film line running under steady conditions, the air inside the bubble will circulate under natural convection currents. The air near the film inner surface rises towards the top of the bubble and falls back down the central axis. The natural convection currents remove some heat from the film near the die but give back to the film the same amount of heat near the top of the bubble. Thus there is no net cooling effect.

4.3.2.1. Heat removal. Internal cooling systems reinforce this natural cooling pattern and also allow the heated air to escape. Air is blown through channels in the die mandrel to emerge radially above the die where it contacts the inner film surface and rises up the inside of the bubble. After cooling the film, the warmed air descends through a vertical tube mounted axially and leaves the bubble through the die mandrel. The cooling air may be directed at the internal surface of the film through a series of discs or through a single diffuser. It is claimed that the latter system gives greater bubble stability and if placed close to the die exit lips will give cooling unaffected by film bubble diameter. Such a system is shown in Fig. 10.

4.3.2.2. Bubble sizing. Since the volume of air inside the bubble alters the blow-up ratio and film width, internal bubble cooling requires that this volume is controlled. This is normally achieved using a sizing basket which can also be seen in Fig. 11. The bubble is contacted above the freeze line by an annular sleeve of wire hoops on which PTFE rolls are loosely fitted to reduce friction on the bubble. The diameter of the sleeve is set to a size appropriate to the layflat width required. One of the hoops pivots from the sizing cage and is used to sense the position of the film bubble. The deviation of the sensing arm from its central position is used to feed an error signal back to a valve positioned in the inlet line of the internal bubble cooling. If the film diameter falls, the valve is opened allowing more air into the bubble whereas when the bubble diameter is above size the signal closes the valve reducing the air flow into the bubble. The design of the sensing arm and control valve should be such

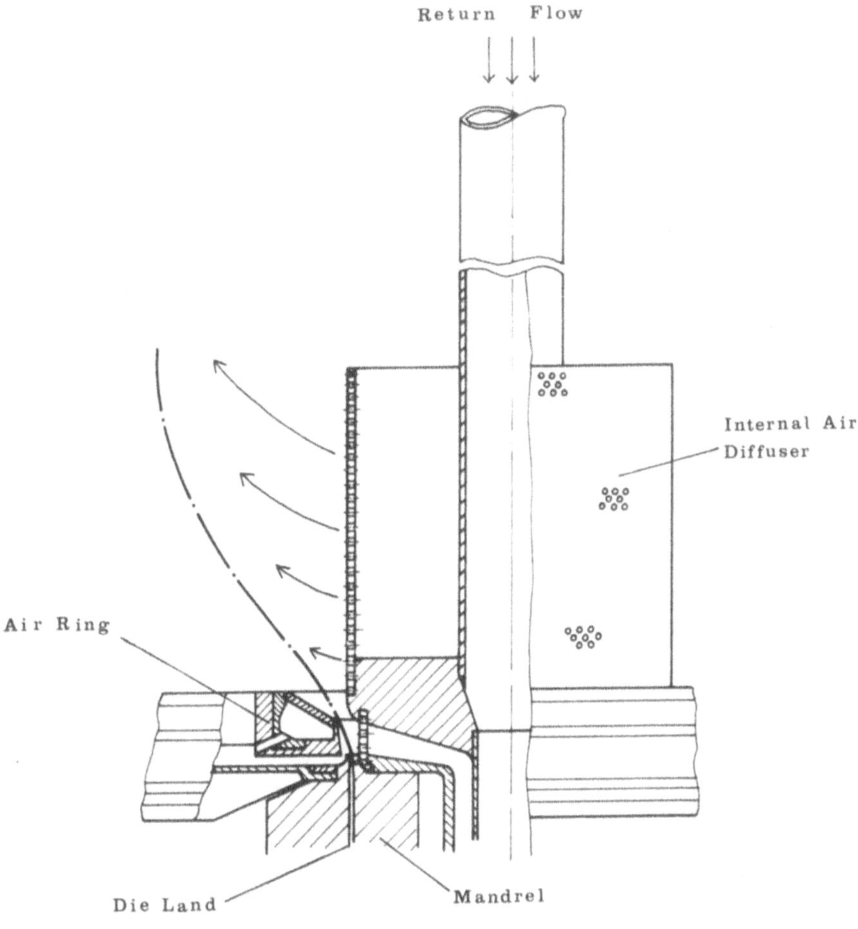

FIG. 10. Internal and external bubble cooling. Based on Reifenhauser design.

that there is a rapid and linear response. The pre-selected load between sensing arm and bubble should be light and constant over the entire range of movement. The sizing basket/cooling intake feedback loop controls the film width to the pre-selected value. Typically layflat width can be kept constant to within 1 or 2 mm.

Whilst, at first sight, it might be thought that the presence of the internal cooling system above the die would make start-up difficult, in practice once the cooling air flow is in operation the newly formed film is easily lifted up the central tube and attached to a line when it reaches the top of the tube.

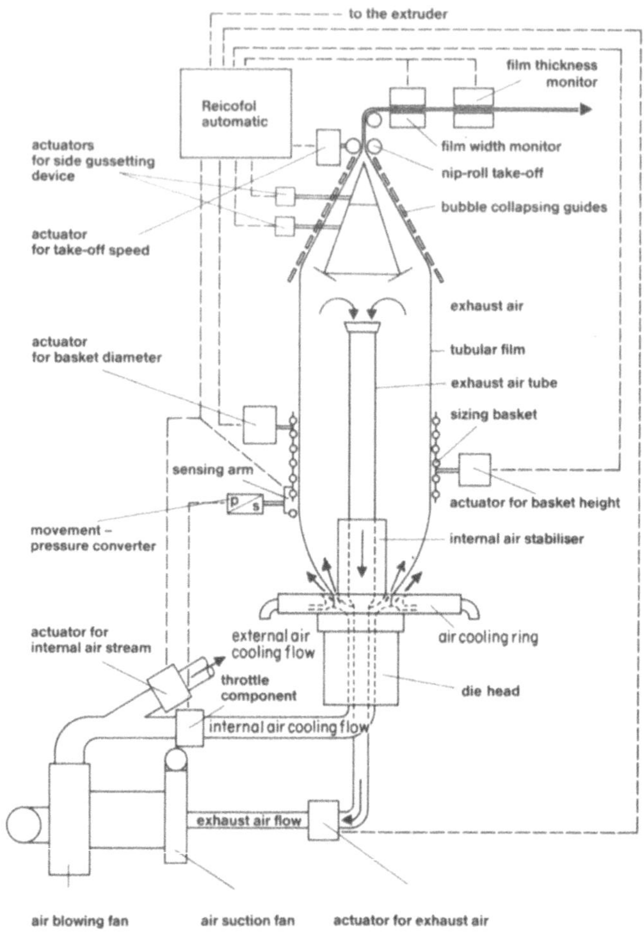

FIG. 11. Automatically controlled line: Reifenhauser IDC system.

Internal cooling is now offered by a number of manufacturers and is widely used. With LDPE films internal cooling allows outputs up to 70% higher than with external cooling. Overall on modern plant, the die diameter (cm) can be multiplied by a factor of up to 12 to give maximum film output in kg/h.

4.4. Film Handling

4.4.1. Collapsing of the Bubble

In the collapse of the film to layflat form at the nip, the need for versatility, to deal with different film materials and dimensions, has encouraged the construction of collapsing frame and nip-roll supports which can be raised or lowered by motorised control. The temperature of the film in the collapsing stage is thus adjustable to suit conditions. Additionally the angle of each collapsing surface may be adjustable and controlled by pushbuttons.

4.4.2. Film Surface Treatment

In order to improve the adhesion of printing inks to polyethylene film, the film surface is treated by passing the film through an electrical discharge. Problems have occurred in this process due to the unreliability of the high voltage supply, danger from high voltages, corrosion of rollers and uneven treatment. These problems have been reduced by use of solid state power supplies and the positioning of a dielectric layer (e.g. quartz) between the two electrodes of the treater. The dielectric coating may be on the surface of a roller over which the film passes. An alternative system involves covering the high voltage electrode with a quartz coating, thus eliminating the need for roll coating and reducing the danger to the operator from electric shock. In the more recent coating units, ozone generated by the discharge is removed through a duct, hence giving less corrosion.

4.4.3. Alternatives to Winding

Some manufacturers have avoided the problems of roll winding at high speed by use of in-line bag making equipment. Alternatively, where film is for internal factory use it may be zig-zag folded into a sided pallet rather than wound on to a reel. This could be especially useful with gusseted film which, because of its variable thickness, presents winding difficulties.

4.5. Automatic Process Control

4.5.1. Process Control by Microprocessor

The advent of the microprocessor has enabled manufacturers to offer film lines where the process is largely automatically controlled during start-up, product change and continuous running. The basis of such control is to monitor various process parameters and use them with the help of computations carried out in a microprocessor to control plant variables so that the film produced matches its specified dimensions of width, thickness and side-gusset size.

4.5.1.1. Control arrangements. Let us consider the control arrangements for an automatic line:

(i) The film width is controlled using internal bubble cooling and a sizing basket. The feedback mechanism described in section 4.3.2 ensures that the bubble is kept to a diameter set by the sizing basket diameter. This is a rapid acting control loop (not requiring the microprocessor) and should be able to cope quickly with holes in the film or other short-term variations. The overall control of width, however, is from the sizing basket diameter and this is set by a signal from the microprocessor.

(ii) The height of the sizing basket, which should be such that the sensing arm is just above the freeze line, is controlled by motorised movement on a signal from the microprocessor. Its position during start-up may be different from that during running.

(iii) The average film thickness is controlled by the nip-roll speed which is set by a signal from the microprocessor. Initially, the microprocessor sets the speed from a knowledge of extruder screw speed and target film width and thickness. However, once the process is in operation, monitoring by an infrared (IR) thickness gauge feeds the microprocessor with information to calculate actual average film thickness. This enables it to initiate appropriate corrective action if necessary.

(iv) The depth of side gussets is controlled by independent motorised movement of two gusseting plates acting on signals from the microprocessor. It is fed information on required depths of gusset and a signal monitoring actual depths.

(v) For overall production control, information on operating conditions, film dimensions, production rates, number of reels produced, etc., can be fed from one or more microprocessor line control units to a central control and display unit. The centralised unit can be used to supervise the running of a large number of lines including setting target dimensions for each line.

Variations and development from the above basic control mechanism are possible and have differed between the systems currently available. These include the Integrated Dimensional Control System (IDC) manufactured by Reifenhauser, the NC–E system by Windmoller and Holscher and a system offered by Gloenco.

4.5.1.2. The IDC system. Figure 11 is a schematic of the IDC system showing the main control paths. Its overall strategy is to feed in target dimensions and some process parameters (e.g. screw speed, blower fan rate, collapsing frame angle and nip-roll height) and then allow the microprocessor to control nip speed, sizing basket diameter and height and side-gusset plate positions from measurements made on the product.

The layflat width is monitored downstream of the nip-rolls using two IR detectors to sense the position of each film edge and thus give accurate film width even with a non-central web. The measured width is used to control sizing basket diameter and thus allowance is made for shrinkage.

An IR thickness gauge based on radiation transmitted through both layers of the layflat film is also positioned after the nip-roll on a carriage which enables it to traverse the film width. Traverse rate is about 2 m/min. A rapid scan of the whole film can be made to determine average thickness. Alternatively the sensor can be set close to the edge of the film where if the die is oscillating a scan of the thickness distribution over the film circumference is produced. Integration of this signal over a complete cycle enables average thickness to be calculated. The thickness gauge also locates the position of the side gussets in the film.

Other features of the IDC system include a linear response function of the internal bubble cooling flow and a rotating sizing basket to smear out any thickness variations produced by non-uniform bubble contact with the basket.

4.5.1.3. The NC–E system. In the NC–E system, predetermined process parameters are normally fed into the control system on a punched tape. Control of sizing basket diameter and hence film width is from the information on the tape. The film thickness is sensed using an IR backscatter gauge targeted on to the bubble surface before the nip-rolls (Fig. 12). Since this line uses nip-roll oscillation to give even reels, the thickness gauge is mounted before the nip-roll on a carousel. The gauge is able to move around the bubble to determine average film thickness and its transverse distribution. The NC–E strategy is thus to monitor and control film before the nip-rolls.

Two additional features are incorporated. Firstly, there is the facility to maximise production rates and maintain the line automatically at maximum output consistent with maintaining the freeze line at or below a predetermined maximum height above the die. During such optimisation, the microprocessor automatically adjusts screw speed, haul-off speed and cooling air volume and maintains film dimensions.

FIG. 12. Infrared thickness gauge on Windmoeller and Hoelscher automatically controlled line.

Secondly, the information fed to the microprocessor on the film circumferential thickness distribution is used to control a sector-cooled die in order to improve film gauge uniformity. The principle of operation is to divide the die into 32 sectors around its circumference. The temperature of each sector can be modified by adjusting the flow of a compressed air stream separately fed to each sector. Lowering the temperature of a sector relative to the rest of the die slightly reduces film melt temperature from that sector. The consequent stiffer melt is stretched sideways less than average thus increasing film thickness in the sector. Conversely, raising the temperature of a sector reduces corresponding film thickness. The NC–E system will adjust each sector according to information from the IR thickness gauge to give a claimed reduction in thickness variability to around $\pm 5\%$ of average film thickness.

4.5.1.4. The Gloenco system. Gloenco offers an automated blown film plant in which the microprocessor is based on the Eurotherm equipment. Width and average thickness are controlled together with a third loop designed to bring the plant automatically up to maximum output rate as judged by a film temperature measurement made just before the nip-rolls.

4.5.1.5. Advantages. The main advantages of automatic microprocessor control are:

(a) more consistent and better control of dimensions,
(b) rapid start-up, and
(c) easy and rapid change between products.

Thus such a control system offers considerable advantage to a processor making a wide range of products from one line and adds to the inherent flexibility of the blown film process.

4.5.2. Lower Cost Control Systems

Parts of the blown film process can be controlled without involving the cost of a fully automatic line. A simple system for width control is shown in Fig. 13. The film width is monitored using a sensing arm attached to a sizing basket and error signals are used to inflate or deflate the film bubble through the normal bubble inflation air inlet pipe. The manufacturers, Sussex Instruments, claim that such a system will bring the bubble width to within 1 mm of the digitally preset value. The same company also markets an add-on-unit for automatic control of average film thickness. The unit monitors film thickness using a pneumatic sensor and controls nip-roll speed from an averaged thickness reading.

For thickness monitoring, display and control, other systems are available as add-on-units to film lines. These are based on either capacitance or β-ray thickness gauging. It is also possible to fit the Kundig IR width control system to an existing film line.

5. RELATION BETWEEN PROCESS AND PROPERTIES

When producing blown film from a given plant, there are a number of process conditions that can be selected to achieve optimum properties in the product. The production of the desired film dimensions, i.e. width and thickness, has already been dealt with. However, some control can be exercised on properties other than dimensions.

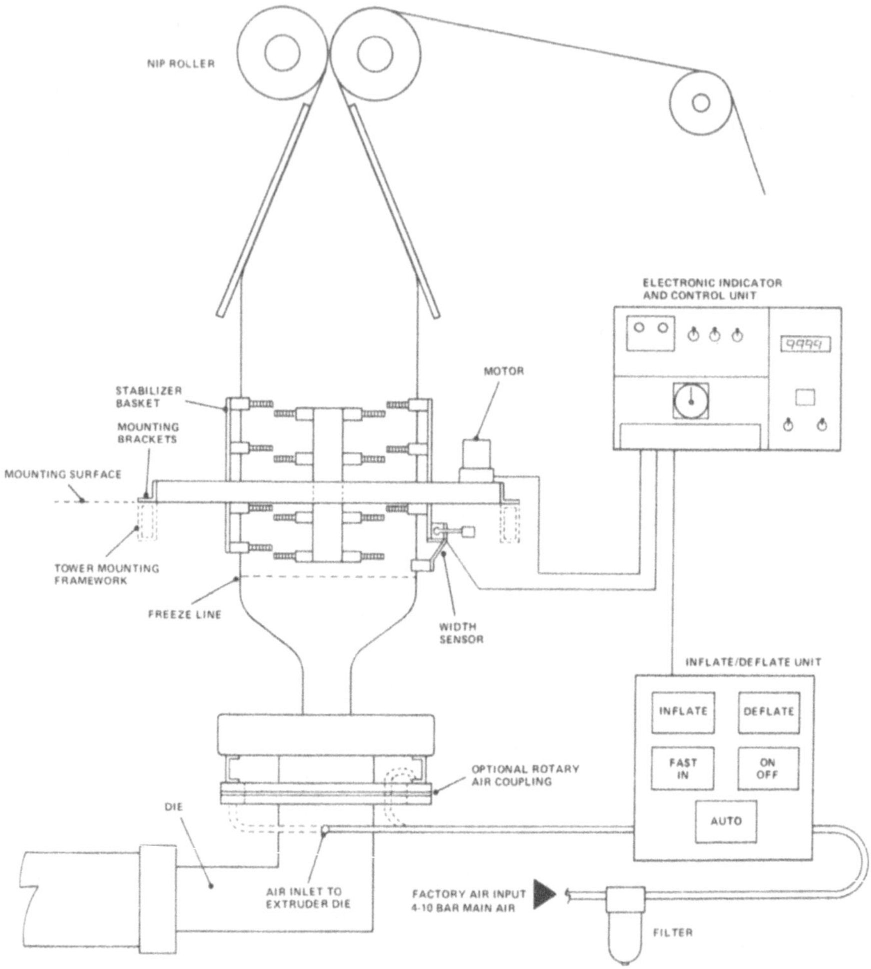

FIG. 13. Diagrammatic view of width control system (not to scale). Sussex Instruments Ltd, Newhaven, UK.

5.1. Commercial Limitations

A number of practical investigations have been reported concerning the effect of changes in process conditions on mechanical, shrink and optical properties of varying grades of polyethylene films. These studies have been useful in understanding processing effects but in general, commercially, a processor is limited in his options in terms of processing variables.

For instance, with regard to output rate, the commercial plant is operated at as high a rate as possible for cost reasons. The output limit is often set by the need to prevent the two sides of the film blocking in the top nips or by the onset of bubble instability and wide gauge variations. For HDPE and LLDPE it is also necessary to prevent excessive shear rates during extrusion. The producer will be operating close to these limits and studies at low output rates are not directly useful.

With regard to blow-up ratio, the layflat width being determined by customer requirements, the choice of die-lip diameter is the deciding factor. Generally a die can be selected to give a blow-up ratio within a given range but the manufacturer will not be free to make small changes in blow-up ratio once the line is operating. Thus for heavy duty LDPE sacks the die size will be selected to give a blow-up ratio in the range 1·5 to 2·0, for general purpose LDPE film a ratio between 2 and 3 and for HDPE a value over 3.

Melt temperature can be altered by choice of extruder operating conditions. However, limits are imposed on the choice of melt temperature. Too low a temperature will give rise to high pressures and torques and possibly unmelted polymer. The upper temperature is set by film blocking and bubble instability. Higher temperatures can be achieved but at the expense of output rate.

The bubble shape and freeze line height are controlled by the air cooling. For a given plastic output rate, though, there are limits set on the one hand to extremely rapid cooling by high air velocities causing bubble deformation and fluttering and on the other hand to very slow cooling by blocking and swaying of the bubble.

Despite the above restrictions a study of the behaviour of the molten and solidifying film between die and freeze line is useful and of interest. This together with its relation to polymer rheology is covered in section 5.2. Quantitative predictions of resulting film properties have not been very successful but an empirical picture with qualitative relationships has been developed and is outlined in Sections 5.3 and 5.4.

5.2. Film Formation

5.2.1. Transverse and Machine Direction Extension Ratios

Between die and freeze line, molten plastic is stretched in both the machine direction (MD) and the transverse direction (TD), i.e. there is biaxial orientation of the melt. There is probably little or no shear deformation in this process.

Ignoring die swell effects, the total TD stretching measured by its extension ratio, λ_T, is equal to the blow-up ratio (BUR). (Note $\lambda = 1 + e$ where e is the strain.) The MD extension ratio, λ_L, can be calculated from eqn. 5.

$$\lambda_T \lambda_L = t_D / t_F \tag{5}$$

Here t_D is the die gap width and t_F the final film thickness. λ_L, sometimes called the draw ratio, is also equal to the ratio of MD velocity at the nip-roll to its value at the die exit. Values of λ_L and λ_T give an indication of the relative amount of stretching in the two directions. In most production processes λ_L is greater than λ_T.

5.2.2. Extension and Temperature Profiles

It is important to note, however, that the longitudinal and transverse stretching do not necessarily take place simultaneously. The bubble shape will determine the extension profiles between die and freeze line. It has generally been assumed that for a bubble with a long neck and late sideways expansion (sometimes called mushroom shaped) the MD extension occurs much sooner than the TD extension. On the other hand, for a bubble whose diameter increases evenly from die to freeze line, the MD and TD extensions occur more nearly simultaneously.

Now, the properties of the molten film will depend on its temperature, which falls between die and freeze line. The temperature profile has been measured by a number of workers using IR radiation thermometers. A recent paper by Wagner[30] shows that there is a drop in film melt temperature which is almost linear when plotted as a function of height above the die face.

5.2.3. Strain Relaxation

A knowledge of extension and temperature profiles is necessary to predict residual strain in the film at the freeze line. Because of relaxation, strains imposed near the die will be reduced by the time the film reaches the freeze line more than strains imposed just before the freeze line is reached. This effect is reinforced by the temperature profile as relaxation will be more rapid at higher temperatures near the die.

5.2.4. Measurement of Melt Extension

Measurement of strain development in the bubble can be carried out by two methods. An approximate result can be obtained at low cost by the first of these.

5.2.4.1. Approximate method. The steady production process is halted by simultaneously stopping nip-rolls and extruder and allowing the film between die and freeze line to cool and so preserve its shape. From the solidified bubble, measurements of circumference and thickness, t, are recorded as a function of x, the distance from the die measured along the film. Figure 14 shows a plot of the calculated bubble radius, r, and t as a

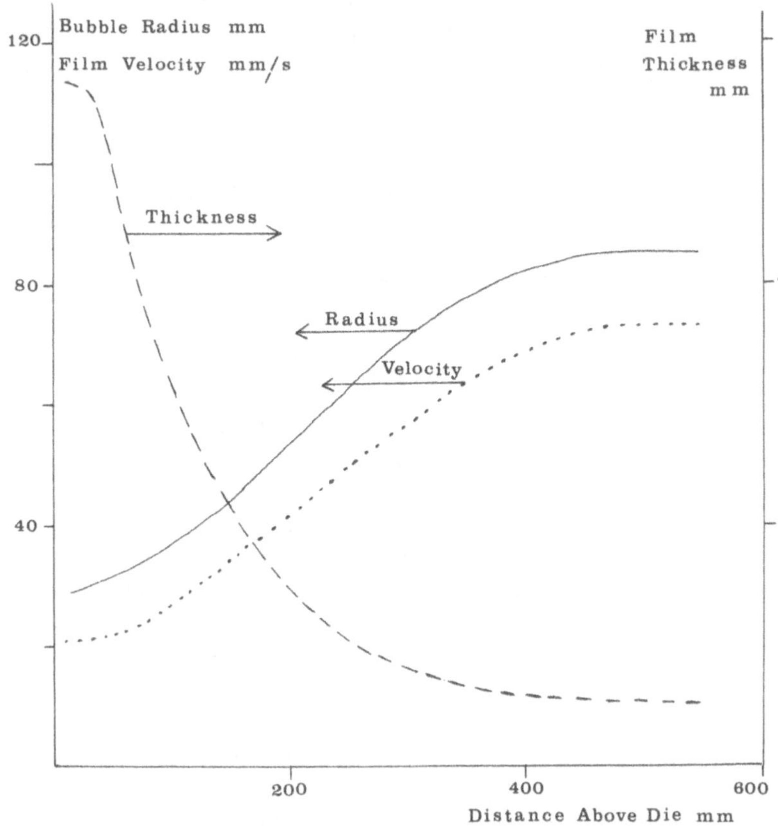

FIG. 14. Bubble dimensions.

function of x. Results were from a film bubble produced from a $2\frac{1}{2}$ in spiral mandrel die with LDPE (MFI = 2·0) in the author's laboratory. The conservation of the mass flow rate, $2\pi\rho rtv$, through the bubble enables the velocity profile to be calculated from r and t values and the haul-off speed (v is the MD velocity and ρ the film density).

This simple method is subject to error due to film dimensional changes in nip-roll slow-down and during its solidification when crystallisation, relaxation of strain and thermal contraction occur. Errors are likely to be most significant close to the die where cooling is slower and die-induced orientation a maximum.

5.2.4.2. Photographic method. A second approach, overcoming these errors, uses cinematography of the bubble on whose surface a very small marker has been placed. The photographs are used to measure radius and velocity profiles and thickness calculated from the conservation of mass equation. Geometric calculations enable results to be plotted as a function of either vertical height above the die or x.

5.2.4.3. Results. Calculation of strain rates in the MD and TD is by use of eqns. (6) and (7).

$$\text{MD strain rate} = \frac{\mathrm{d}v}{\mathrm{d}x} \tag{6}$$

$$\text{TD strain rate} = \frac{v}{r}\frac{\mathrm{d}r}{\mathrm{d}x} \tag{7}$$

Figure 15 shows the strain rate profiles calculated from the author's results. More accurate measurements of film changes between die and freeze line using cinematography have been reported.[26,30-33]

These results show an 'S' shaped curve for bubble radius against height above die, with TD strain rates passing through a maximum just before the freeze line is reached. In some cases, the MD strain rate decreases steadily towards the freeze line from values measured closest to the die and in other cases shows a maximum as in Fig. 15. This MD maximum, when it occurs, is found before the TD maximum.

5.2.5. Residual Strain, Crystallisation and Film Properties

Together with temperature measurements, the strain rate profiles suggest that MD strains should relax more than the TD strains before reaching the freeze line. As a result residual strains in the two directions could be similar depending on relative total values and details of strain profiles. Estimates of residual strains, taking into account melt relaxation, were compared to the shrinkage properties of the resulting films by Farber and Dealy.[32] Correlation was disappointingly poor. It must be concluded that orientation in the solid film also depends on the crystallisation processes occurring at the freeze line.

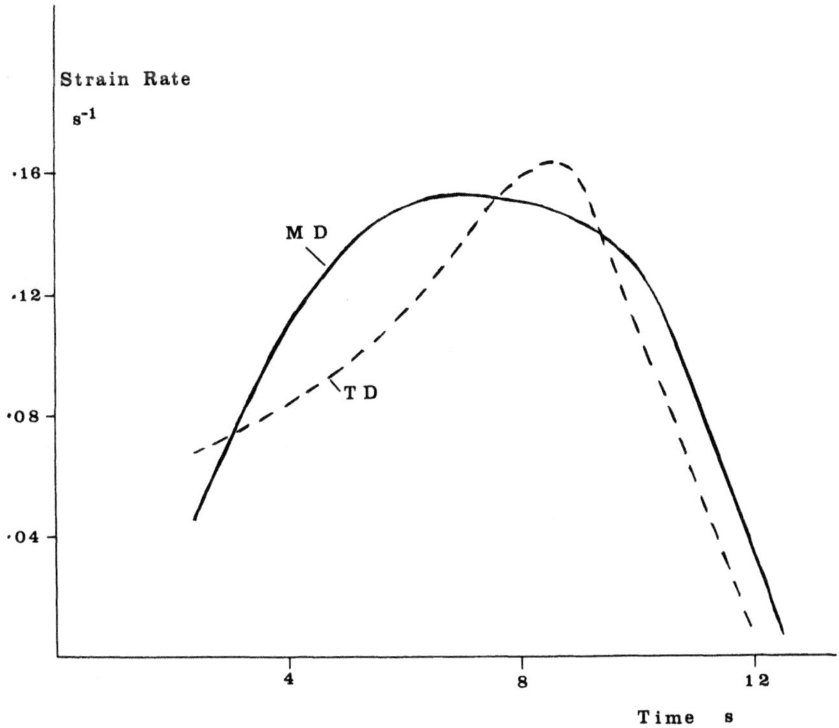

FIG. 15. Melt strain rates in bubble as a function of time since leaving die for LDPE.

A fuller investigation of crystallisation in blown film was reported by Holmes and Palmer[34] and more recently by Nagasawa *et al.*[35] The latter measured changes in film birefringence below and above the freeze line whilst working with a BUR of 1. Birefringence, and by inference orientation, increased to the greatest extent after the freeze line as crystallinity developed. Nagasawa and colleagues proposed a detailed mechanism for the crystallisation process to explain film morphology which they studied by microscopy and X-ray diffraction.

It may be concluded that film structure and hence properties are influenced by the rheological behaviour of the melt below the freeze line and by the complex crystallisation process. Details of these interrelations have not been fully investigated.

5.2.6. Prediction of Bubble Shapes

To conclude this section it is worth considering attempts to predict bubble shapes and melt behaviour from more basic rheological knowledge of the plastic melt. Difficulties arise firstly due to a lack of basic rheological data on stress, strain and strain rate relationships for polymer melts undergoing extensional deformation and secondly due to the non-isothermal nature of film blowing. Nevertheless some progress has been made.

Petrie[36] summarises the theoretical work undertaken by himself and Pearson and compares this with published experimental results. When bubble radius profiles are calculated from models assuming either a non-isothermal Newtonian melt or an elastic melt, they fit the experimental data reasonably well. However when compared with strain rate data, agreement is not very good. Han and Park[33] propose a melt represented by a power law flow equation. To obtain the basic parameters for their model, they studied isothermal film blowing using a heated chamber above the die. When applied to the conventional blowing process their results gave good agreement with experimentally determined bubble radius and film thickness profiles.

None of the models includes the influence of the relaxation of shear-flow induced strain from the die (die swell), nor are their basic rheological data taken completely from independent (non-film blowing) measurements. Thus at all levels of investigation, the process of film formation is not properly understood. Nevertheless, useful empirical relationships between film properties and processing conditions have been found and are covered in the following sections.

5.3. Shrink Properties of Polyethylene Film

Film shrinkage on heating is arguably the property most likely to be clearly dependent on the melt strain behaviour. Whilst as stated above a quantitative dependence has not been established, shrink behaviour can be qualitatively related to processing conditions during film formation.[5,37] For the purpose of establishing these relationships, shrinkage is determined in both MD and TD by heating a film to 115°C and measuring the percentage reduction in the distance between marks placed on the film.

The most significant factor affecting shrinkage is the BUR. Increasing BUR increases shrinkage in the TD and reduces it in the MD. At low BUR, MD shrinkage is greater than TD; e.g. at a BUR of 1·5, the MD shrinkage can be 80% with TD shrinkage at 25%. However, at a BUR of about 4, a film with balanced shrinkage of about 60% in both directions can be produced. Bubble shape also affects shrinkage. A bubble with a narrow neck near to the die and late sideways expansion will tend to produce a more balanced film than one where the sideways expansion is close to the die. Additionally there is some evidence that increasing throughput rate whilst maintaining freeze line height so that the rates of drawing are increased in both directions will give higher shrinkage in the film. All of the above trends fit the picture of melt strain development and relaxation covered in section 5.2.

Melt temperature and freeze line height do not influence shrink behaviour directly. They may however modify bubble shape. Polymer molecular weight has an influence with high molecular weight (low MFI) polymers showing greater shrinkage. The use of these materials (MFI 0·2–0·5 g/10 min) is common for pallet load shrink wrapping, but, because of better optical properties, LDPE of MFI 0·7–2·5 g/10 min is often preferred for the retail market.

5.4. Film Strength Properties

5.4.1. *Evaluation*

There are three main tests used to assess film strength properties. (a) Tensile tests (ASTM D882; ISO R1184; BS 2782 Method 326), using dumb-bell shaped samples cut so that the tensile stress is applied either in the MD or TD, are used to assess the ability of the film to sustain static loads. (b) Tear strength can be assessed in both directions (MD tear strength refers to results from a test in which the tear propagates in the MD). Elmendorf (ASTM D1922) or trouser type (ASTM D1938) tests can be undertaken. Results are often variable and difficult to interpret. (c) One of the most useful tests for correlating with performance has been the falling dart impact test (ASTM D1709–62T, BS 2782 Method 352). The drawbacks to this test are the large number of samples required, difficulty in ensuring consistent samples and the variable dart velocity as it is stopped by the film. A modified test has been incorporated in DIN 53373 and developed by DSM in Holland. The general arrangement of the method is as in the conventional dart test, but the dart's mass and energy on hitting the film are very much higher so that failure will always occur. By fitting an electrical accelerometer in the head of the dart, the energy loss to cause film failure is measured for each test. Results generally correlate with standard tests but give lower values for HDPE and PP films, where impact strength depends strongly on test speed.

5.4.2. *Results for LDPE*

The relationships between film strength and processing conditions were first investigated for LDPE (for example refs. 37, 38). Strengths appear to be dependent on both the inferred level and balance of orientation in the film. A brief summary of the broad relationships can be made.

Generally tensile strength (or yield strength) increases with increasing orientation in the direction of the tensile force. For high MD tensile properties, a low BUR giving nearly monaxially oriented film is required.

High tear strength values are found in those films where the level of orientation is low. An example could be a thick sack film made at relatively low BUR. Whilst TD tear strengths are not strongly dependent on processing conditions, the MD values appear to be reduced by those factors which increase impact strength.

Improvement in impact strength is achieved by high balanced orientation of the film. This can be achieved by increasing the BUR, by using a suitable bubble shape in order to ensure that TD orientation occurs after MD orientation and, generally, by increasing output rate. Melt temperature also influences impact strength somewhat but the reported patterns of behaviour are not consistent with each other.

5.4.3. Results for HDPE

When the manufacture of HDPE films was attempted it was readily apparent that films made with a conventional LDPE bubble shape were unsuitable having low impact strengths and a tendency to split. Using LDPE experience, higher BURs were tried for HDPE films together with a long neck giving sideways expansion close to the freeze line, which was kept high. Additionally, high molecular weight polymers were used. These measures gave an acceptable product.

Early reported studies[11,39,40] gave results which were somewhat confusing, especially when compared with industrial practice. For instance Davies and Gray's studies[11] showed that increasing BUR in the range 2–5 did not improve impact strength but gave a more balanced film by reducing MD strength properties. Paschke *et al.*[40] found that strength results for thin films could be higher than for thick films especially if expressed as strength per unit thickness.

More recently the picture appears clearer.[41] Die-induced orientation relaxes during the passage of the melt up the neck to just below the freeze line where TD and MD orientation occurs as the bubble expands and thins. It appears that the rate of stretching at this stage is significant, with rapid stretching giving the best impact results. Thus throughput rate is an important factor.

In a production situation, for a given film width, the die diameter determines the BUR. If the diameter is high, a low BUR will ensue with reduced impact strength due to an unbalanced film. If the die diameter is too low, the BUR will be high but the output rate will be reduced as the high molecular weight HDPE has a high viscosity which limits the shear rate attainable. The low output rate implies a low haul-off speed and the consequent low stretching rate gives film with reduced strength. Thus there is an optimum BUR which is often around 4.

Empirical knowledge thus enables film strengths to be related to processing conditions in a qualitative manner but a full understanding is still lacking.

5.5. Optical Properties

5.5.1. Haze and Gloss

Most packaging applications involve close contact between the packaging film and its contents. This means that the most useful measures of film optical performance are haze and gloss: distortion is relatively unimportant.

Haze is a measure of 'cloudiness' and is determined from the percentage of transmitted light scattered by more than a given angle (usually 2·5°) on its passage into and through the film (ASTM D1003; BS 2782 515A). Where the contents of a package need to be seen, low haze is desirable.

Gloss measures the amount of mirror-like light reflection from a film. The light reflected at the same angle to the film as the incident light is measured. Normally the angle of incidence selected is 45° (ASTM D2457; BS 2782 515B), and the gloss expressed in terms of gloss units which are numerically 1000 times the ratio between reflected and incident light intensities. High gloss is thought to be desirable in packaging films.

5.5.2. Sources of Haze

The majority of interest has concerned the optical properties of LDPE films. Early studies[31] established that light scattering (haze) occurs mostly from surface scattering in the case of most clear packaging films (thickness less than say 40 μm). The internal contribution to scattering is relatively small but obviously increases in importance with increasing film thickness.

It was also established that the scattering of light at the surface of the film is due to two distinct causes. Firstly, the surface of the melt as it emerges from the die is not smooth but has a fine scale roughness, due probably to the elastic nature of the melt as it enters, traverses and leaves the die land. This 'extrusion' roughness decreases between the die lips and the freeze line due to surface tension and stretching of the molten surface. Secondly, crystallisation of the polyethylene which occurs near the freeze line gives rise to entities which scatter incident light. The amount of crystallisation and the nature of the crystallites formed will determine how much light is scattered. It is expected that the cooling rate of the melt will influence crystallisation, with slow cooling giving rise to more crystallisation.

5.5.3. Effect of Processing Conditions

This understanding of the sources of haze led to the prediction of the way in which processing conditions should affect optical properties. For example, if film is made under constant conditions with the exception of the cooling air rate then, starting at high cooling rate with consequent low freeze line, it is possible progressively to reduce cooling and raise the freeze line distance. Figure 16 shows that haze is expected initially to decrease due to fewer extrusion surface defects and then increase due to increased crystallisation. For some grades of polyethylene and under appropriate conditions such a variation has been demonstrated.

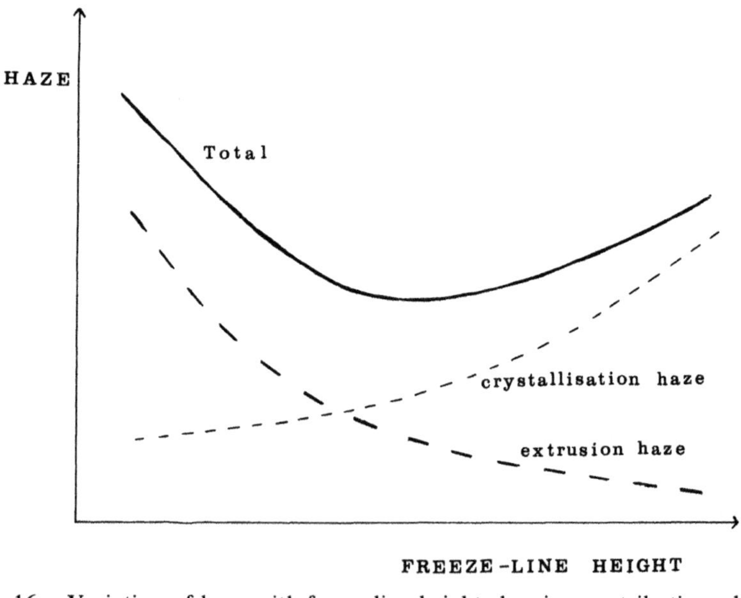

FIG. 16. Variation of haze with freeze line height showing contributions due to extrusion defects and crystallinity in LDPE film.

In general the factors which encourage relaxation of extrusion defects are a high melt temperature, stretching of the film and increased time before solidification. Factors which reduce the haze from crystallisation are not so clearly understood but include rapid cooling.

Later studies[12,37,42,43] have clearly shown that the relationship between optical properties and processing is complex and varies according to the grade of polyethylene studied. The use of lowest density grades (e.g. 915 kg/m^3) emphasises the importance of extrusion defects (which depend on die design and detailed rheological behaviour) whilst properties of film from grades of a higher density (e.g. 927 kg/m^3) will be more dependent on crystallisation conditions. In some cases high levels of additives will raise light scattering and mask effects due to processing.

Whilst the picture is complex, the following relationships appear to be demonstrated:

(a) A melt temperature increase will improve optical properties.
(b) For the lowest density grades of LDPE, a high freeze line will give the best optical properties whilst for higher density grades the freeze line height would need to be reduced.
(c) A blow-up ratio between about 2 and 3 will give the best optical properties but this depends on the polyethylene grade.
(d) Increased output tends to give worse optical properties.
(e) Selection of high MFI grades will give better optical properties.

6. MATERIALS FOR BLOWN FILM

6.1. Low Density Polyethylene (LDPE)

6.1.1. Manufacture and Structure

Low density polyethylene is manufactured from ethylene by a high pressure process in which polymerisation is started by an initiator. The nature of the manufacturing process causes branching of the polyethylene molecules. Both short and long chain branching occur. The short branches to the main chain inhibit the tendency of the polyethylene to crystallise so that LDPE has lower crystallinity and, hence, lower density than a more perfect less branched polyethylene material. Density can be controlled in the range 915 to 930 kg/m^3 by varying polymerisation conditions.

The long chain branching of the LDPE molecules is expected to modify their flow properties. This is because a branched molecule is expected to be more compact than an unbranched molecule of the same molecular weight. However, investigations into the effect of long chain branching on rheology have often given contradictory results.[24] This confusion is probably associated with the wide variability (or distribution) of both molecular weight and branching in any sample of LDPE.

6.1.2. *Molecular Weight*

Producers of LDPE supply material in a range of grades of differing average molecular weight. An indication of the relative molecular weight of a polyethylene grade is given by the MFI of the grade. This index measures the flow rate of the molten polymer under standard low shear rate conditions (ASTM D1238, ISO R292, BS 2782–720A, DIN 53735). Lower values of melt index correspond to grades with higher average molecular weight and consequent higher melt viscosities. Most film grades of LDPE have an MFI between 0·5 and 4·0.

The distribution of molecular weights of a polyethylene will also be important in determining its film-forming properties. Reducing the molecular weight distribution will reduce the non-Newtonian character of the melt flow, make the viscosity more strongly dependent on temperature and alter the tendency to melt fracture or 'sharkskin' effects.[24]

6.1.3. *Melt Flow Properties*

Figure 17 shows how the apparent shear stress varies with apparent shear rate for a film-grade (melt index 2·0) LDPE sample at two different melt temperatures. Using a capillary rheometer with a 20:1 L/D die ratio, flow rates were obtained from the ram displacement and shear stresses from the pressure monitored above the die using a strain gauge melt pressure transducer.

Similar curves have been reported by Williamson[37] showing that reducing the melt index raises the shear stress at all shear rates but by a proportionately smaller amount at higher shear rates.

6.1.4. *LDPE Film*

The general suitability of LDPE for conversion to a relatively tough, flexible and semi-transparent film by the tubular blown film process has been utilised for 40 years. Development work has concentrated on how the more detailed molecular characteristics of LDPE materials can affect processing behaviour and film properties.

Williamson[37] reports the effects of varying melt index. Higher melt

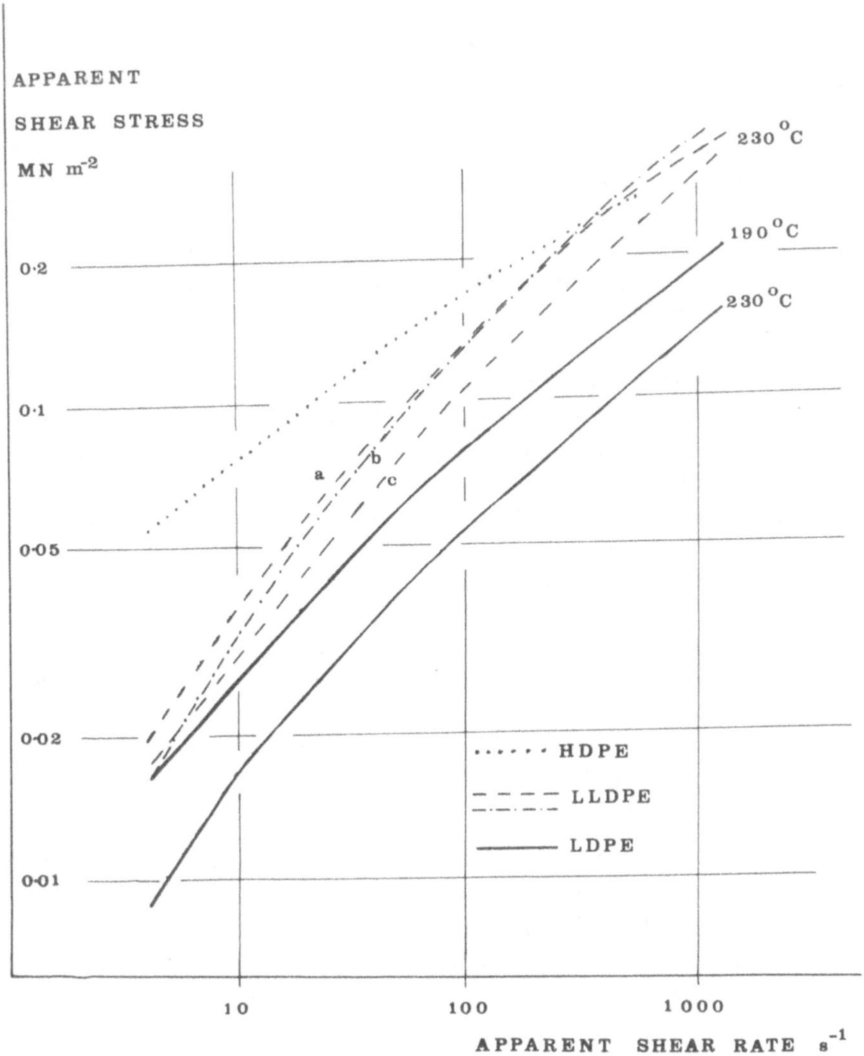

FIG. 17. Flow curves for film-grade polyethylene polymers. HDPE is Hostalen. LLDPE: a, Dowlex 2045; b, News 8001; c, Dowlex 615 00.38. LDPE is Alkathene XJF 46.

index grades (i.e. grades with a lower average molecular weight) at a given flow rate through a die will produce less melt turbulence or distortion of the extrudate surface and hence films with less haze and higher gloss. These grades also enable the melt to be drawn down to produce thinner films at a given melt temperature. However, bubble instability (swaying and susceptibility to draughts) increases with increasing melt index and consequently it is unusual to use a melt index grade higher than 4 in film production.

Mechanical and impact properties of LDPE films depend on the melt strain rate–time–temperature profile during film formation (see section 5.4). The effects of changes in polymer characteristics on these properties are thus complicated. However, it appears that lower melt index grades give films with better impact properties. This may be due partly to the higher viscosity melt retaining a greater degree of molecular orientation than a low viscosity melt and partly to an intrinsic improvement in impact properties with an increase in molecular weight. Low melt index grades will also give greater shrinkage for shrink film applications. They are used to make thicker films in applications where optical properties are not critical.

The effect of molecular weight distribution has been examined by Perron and Lederman.[44] They found that increasing the molecular weight distribution (as measured by an increase in M_w/M_n ratio) led to films with increased haze and less gloss but with higher dart impact strengths. The change in optical properties was associated with an increase in film surface irregularities with an increasing molecular weight distribution.

Explanations of these observed effects are difficult and perhaps meaningless without full characterisation of all of the following in a coordinated study:

(a) The polyethylenes, including their long chain branching and molecular weight distribution,
(b) their rheological properties including elasticity and die swell,
(c) the orientation and cooling processes during film formation and
(d) the morphology and orientation of the films produced.

6.2. High Density Polyethylene (HDPE)

6.2.1. Manufacture and Structure

By using appropriate catalysts, polyethylene can be produced at lower temperatures and pressures than those used in the manufacture of LDPE. Use of a catalyst means that branching is reduced compared with the high pressure process. The reduction in the number of short chain branches produces polyethylene molecules which because of increased regularity are able to form a solid polymer that has a higher degree of crystallinity than LDPE. This gives a denser product, typically around 960 kg/m^3.

The polymerisation conditions are such that additionally HDPE will have less long chain branching and a generally lower molecular weight distribution than LDPE.

6.2.2. *Properties and Applications*

The level of crystallinity in polyethylene affects both mechanical and optical properties. Compared with LDPE, the modulus or stiffness of HDPE is greater by a factor of about 4, whilst its tensile yield strength at about 25 MN/m^2 is approximately twice that of LDPE. However, its impact strength can be rather worse than that of LDPE, especially at low molecular weights.

Apart from giving a stiffer, crisper film the high level of crystallinity impairs transparency to give a rather opaque blown film. Under normal production conditions the film will have a matt surface with low gloss.

Its mechanical properties are such that HDPE can be used to make quite low gauge (10–25 μm) films used mainly for overwrapping sheets or counter bags. More recently, it has been used for waste disposal bags, e.g. pedal-bin and dust-bin liners. Thicker films are sometimes used to manufacture carrier bags with a matt appearance and feel.

6.2.3. *Film Production*

6.2.3.1. Effect of high molecular weight. It was early realised (e.g. ref. 39) that, for a satisfactory stiffness and impact strength with a matt surface, high molecular weight grades (e.g. melt index 0·2–0·4 at 190°C under 5 kg load) are necessary. Low molecular weight grades (e.g. melt index 1–5 g/10 min at 190°C under 5 kg load) gave longitudinally orientated, glossy surfaced films with low strength and a tendency to split.

With an appropriate bubble shape (section 5.4.3) and suitable collapsing frame positioning (section 3.5.1) satisfactory films can be made, provided the melt temperature is kept sufficiently low. This is difficult due to the high shear stresses involved. Figure 17 includes the flow properties of a typical HDPE film grade, showing its relatively high shear stresses.

To prevent high melt temperatures with these high viscosity materials, extrusion screw speeds must be kept low and special screw designs used (see section 3.1.1). Additionally greater drive motor power is necessary.

6.2.3.2. Molecular structure effects. A start has been made to investigate the effects of molecular weight distribution and long chain branching on HDPE film formation.[45] Different grades of HDPE, of similar average molecular weight and with similar high shear rate viscosities, can have different elongational melt viscosities, melt relaxation times and low shear viscosities due to differences in molecular weight distribution or long chain branching. It is likely that the presence of a high molecular weight tail to the molecular weight distribution or some long chain branching in the polymer will give:

(a) increased elongational viscosity which will become more pronounced on continued stretching of the melt (tension stiffening);
(b) reduced melt elongation at break;
(c) longer melt relaxation times; and
(d) greater non-Newtonian flow behaviour and hence increased low shear rate viscosity.

Such a polymer will give higher pressure drops across the die and, more importantly, greater difficulty in film drawing between die and freeze line. It will show a greater tendency for melt fracture and imperfections in the molten film will be less likely to flatten due to the longer melt relaxation time. Consequently, the film produced will have a poor finish and impact strength.

In contrast, a grade of HDPE with a short relaxation time and a less elastic melt will give relatively strong, melt fracture free film with a high elongation at break.

6.3. Linear Low Density Polyethylene (LLDPE)

6.3.1. Manufacture and Structure

Linear low density polyethylene is a copolymer of ethylene and is produced by similar methods to those used for the manufacture of HDPE, i.e. using catalysts at relatively low pressures and temperatures. Although such copolymers have been available from Dupont for some years, it is only recently that their potential benefits have been promoted. At present, the two main suppliers are: (a) Union Carbide (Unifos in Europe) who use a fluid-bed gas phase polymerisation process and market the LLDPE as a granular powder under the 'News' trademark; and (b) Dow Chemical using a liquid phase solution process and marketing pellets of LLDPE under the trade name 'Dowlex'. The number of suppliers of LLDPE is however growing.

The comonomers used with ethylene in the production of LLDPE are chosen to produce short chain branches on the polyethylene-type backbone. This can be achieved by use of α-polyolefins such as propylene, butylene or hexylene. Increasing the proportion of the comonomers will increase the level of short chain branching and, hence, reduce the level of crystallinity in the resulting polymer. Manufacturers thus have a means of controlling polymer density and to date grades with a range of density from 919 to 935 kg/m^3 have been marketed. The lower end of this range is comparable with LDPE.

However, the polymerisation methods for LLDPE give rise to a material which unlike LDPE has little or no long chain branching and a narrow molecular weight distribution. These structural differences give rise to melt rheological and solid mechanical properties differing from these of LDPE.

Suppliers of PE thus have available in LLDPE the facility of manufacturing a LDPE of density comparable with high pressure PE without the necessity for expensive high pressure polymerisation facilities.

6.3.2. Melt Rheology

Figure 17 includes the flow curves for three LLDPE polymers. These are Dowlex 2045 (MFI 1·0 and density 920 kg/m^3), Dowlex 615 00.38 (MFI 1·0, density 935 kg/m^3) and News 8001 (MFI 1·0 and density 919 kg/m^3). The Figure shows that these LLDPE polymers are more Newtonian than both normal LDPE polymers and HDPE film grades. (As the shear rate increases the shear stress increases more rapidly than in the other PE materials.) This fact means that at the shear rates found in the film die and extruder, the LLDPE will have a relatively high apparent viscosity and consequently will give high back pressures and require high motor torque.

However, at the relatively low elongational shear rates found in the bubble (see section 5.2.4) the LLDPE will probably have a relatively low elongational viscosity. This will give the material the ability to be drawn down easily but may cause problems with bubble stability. The easily deformed molten film could be distorted by excessive blowing air or other disturbances.

6.3.3. Film Properties

Suppliers of LLDPE for film claim that, compared with LDPE, the new materials offer at the same film thickness greater toughness, tear and puncture resistance and strength. Alternatively, the quality of LDPE film can be maintained at reduced film thicknesses.

In examining the mechanical properties of LLDPE it should be remembered that this material can be manufactured in a range of densities. Some properties of LLDPE are associated with higher density. For example, Fig. 18 shows how the yield strength of a range of polyethylene film materials increases as the density of the PE increases. The figure includes results on LDPE, HDPE and LLDPE, showing that LLDPE is in line with the general pattern. A similar behaviour is to be expected with film modulus but here there is some evidence that LLDPE may have a slightly higher modulus than conventional LDPE of the same density.

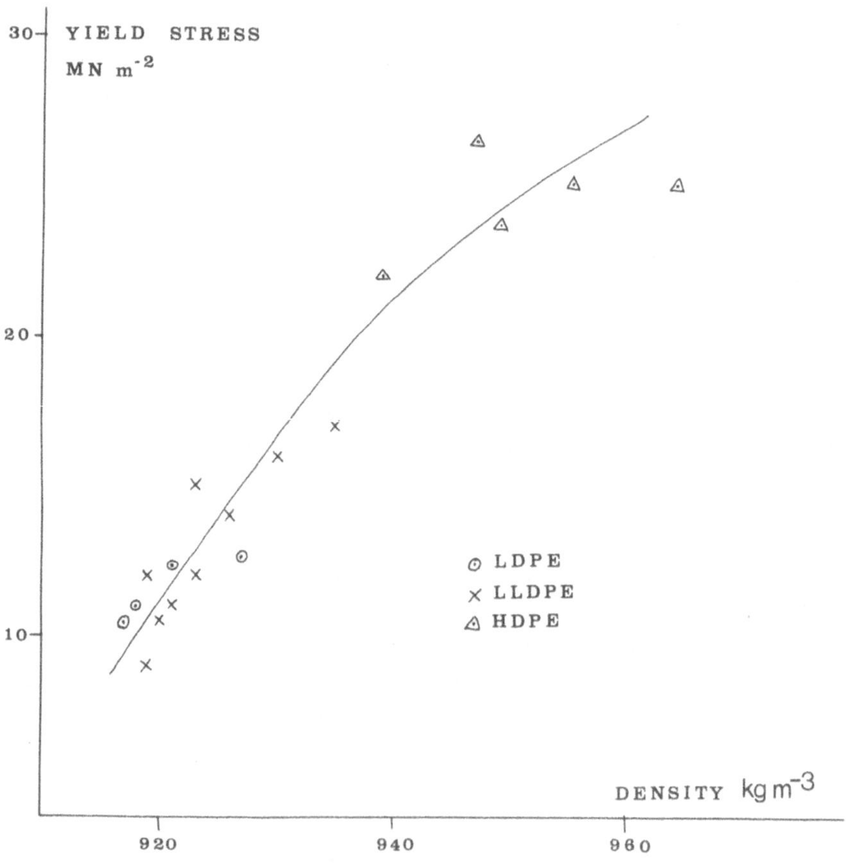

FIG. 18. Tensile yield stresses of different polyethylene films.

The ultimate tensile properties of LLDPE are, however, demonstrably greater than those of LDPE. Figure 19 shows the high elongation at break figures for LLDPE. Correspondingly high tensile strengths are also found. These results indicate that, at least at moderate elongation rates, the energy necessary to break a sample of LLDPE is greater than for the corresponding LDPE. This fact gives rise to increased tear and puncture resistance.

In the case of impact tests, the published results are not so clearly in

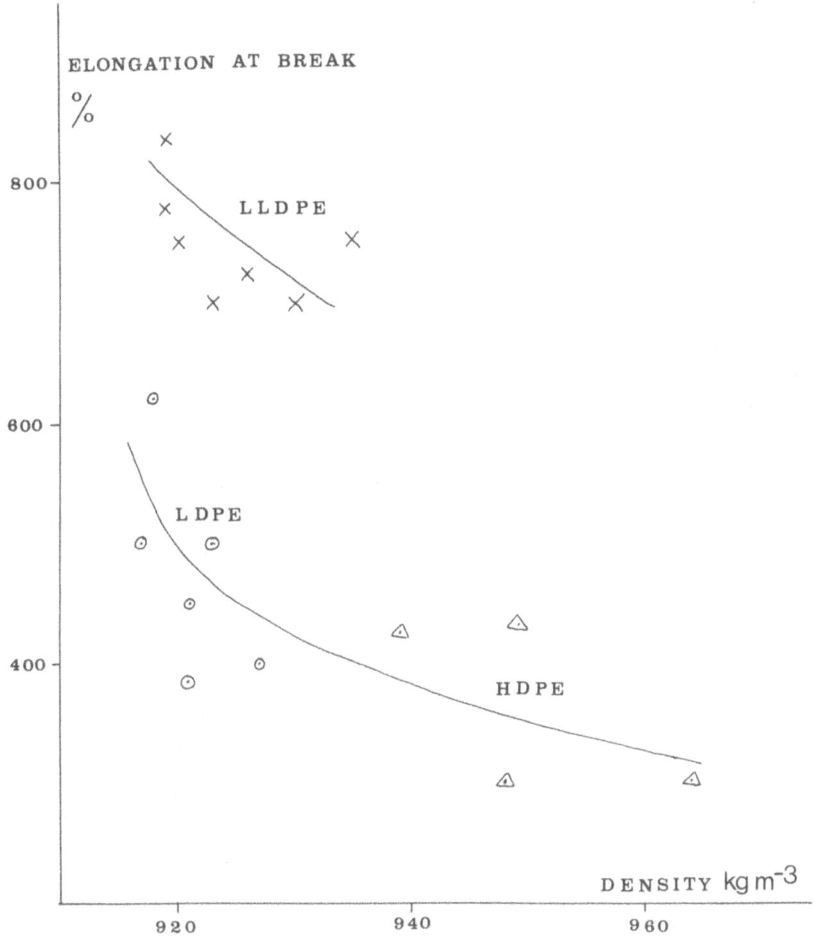

FIG. 19. Variation of tensile elongation at break with polyethylene film density showing high values for linear low density polyethylene (LLDPE).

favour of LLDPE, as can be seen in Fig. 20 showing falling weight dart impact results. However, impact results depend greatly on film manufacturing conditions and are thus less easily compared.

The disadvantages of LLDPE appear to be slightly inferior optical properties compared with LDPE and possibly the need to use additives

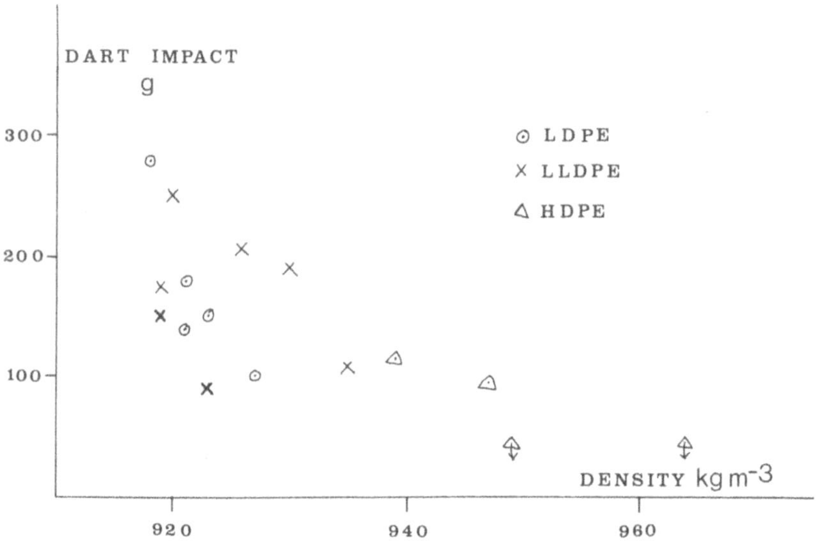

FIG. 20. Dart impact values of 40 μm polyethylene film of different densities.

(e.g. slip, antiblock, antistatic) at about twice the concentration normally found in LDPE. There are also differences in processing behaviour, discussed in the next section.

6.3.4. Manufacture of LLDPE Blown Film

6.3.4.1. Use of unmodified LDPE plant. Linear low density polyethylene may be produced on some unmodified LDPE blown film equipment but at lower output rates (a 30% or more reduction). Outputs of 180 kg/h from a 90 mm extruder have recently been suggested by Jeffries.[46] Difficulties in using unmodified equipment arise because of the rheological behaviour of LLDPE. Its flow properties mean that at the flow rates found in the extruder and die LLDPE has a high apparent viscosity (similar to a high molecular weight HDPE). This can lead to excessive extruder drive torque requirements, excessive heat build-up in the polymer, and very high pressures in the die and melt fracture of the film as it leaves the die lips. However, at low shear rates the shear stresses for LLDPE are not large. In the elongation of the melt between die lips and freeze line elongation rates are low and the LLDPE is easily deformable, behaving like a high melt index LDPE. This can give problems with cooling, preventing the use of high velocity air.

6.3.4.2. Plant for LLDPE. With modified or purpose-built plant, LLDPE film can be produced at comparable output rates to LDPE. It should be noted that an equivalent mass output rate for LLDPE would imply an increase in linear line speed if the film has been down-gauged. Several authors have discussed LLDPE film production.[47–49]

To reduce the melt viscosity, LLDPE is processed at melt temperatures in the range 205–230°C. The actual temperature depends on the material (low MFI grades require higher temperatures), the flow resistance of the die and the cooling system. A higher temperature reduces pressures and torques but may give problems with bubble stability at high output rates.

The design of a die, suitable for LLDPE, is aimed to reduce the die flow resistance and hence reduce pressure build-up, screw torque and polymer overheating. This is achieved by increasing the die gap width to between 1·5 and 3·3 mm and reducing the land length to as little as 10 mm. Other flow passageways within the die may also be increased in size. These changes will reduce the shear rate at the die exit and thus keep it to a range from 100 to 300 s^{-1}, above which melt fracture may occur. It has been found that spiral mandrel dies are the most successful die type (the recombination of melt streams necessary in other designs leads to weak lines in LLDPE film).

The extruder drive for processing LLDPE must be sufficiently robust. Table 4 shows the data developed from the paper by Wright and Knittel[49] indicating motor power requirements.

TABLE 4
EXTRUDER PARAMETERS WHEN PRODUCING LLDPE FILM

Screw diameter (mm)	*Maximum screw speed (rpm)*	*Motor drive power (kW)*	*Maximum output (kg/h)*
30	140	5·5	21
50	120	20	74
70	110	45	170
90	105	90	320
120	100	180	650

To give a controlled melt temperature despite the high shear stresses involved in LLDPE, recommended screw designs have rather deeper channels than for LDPE, a compression ratio of about 2·5 to 1 and preferably contain a mixing section. Screws with barrier flights (as discussed in Section 4.1.2) have worked successfully provided the flight gap is not too tight.

The modifications to extruder and die are aimed to cope with the high shear stresses at the shear rates involved. The relatively low melt strength in the bubble reduces the maximum (sustainable) cooling air velocity. It is therefore necessary to use dual- or multi-orifice cooling rings. Internal bubble cooling will also be helpful.

The total amount of drawing between die lips and freeze line will be high, due to a wide die gap and possibly a thinner than normal film. This drawing is helped by the low extension viscosity. At present, the bubble is run with a freeze line height between 1 and 1·5 die diameters and with a maximum blow-up ratio of about 3.

It has been found that for slitting, hardened steel or heated blades are necessary if they are to last. Web tensions and winding behaviour depend on film stiffness which varies with LLDPE density.

The heat sealing range of LLDPE is higher and somewhat narrower than LDPE. It has been reported that bag making machines have to run at reduced speeds when using the new material.[50]

6.3.5. Future Prospects

It seems likely that the use of LLDPE film will grow at the expense of LDPE film. This is due to the desire of material suppliers to use a low pressure manufacturing process, the somewhat improved properties and the increasing ability with appropriate equipment to produce LLDPE film at economic rates.

At present world production capability for LLDPE is about 500,000 tonnes per annum. O'Neill[51] states that in Western Europe the present capacity is 100,000 tonnes and this is expected to grow to 750,000 tonnes by 1986.

6.4. Modified Polyethylenes

6.4.1. Ethylene/Vinyl-Acetate Copolymers (EVA)

Because of the comparable reactivity of vinyl acetate and ethylene, EVA copolymers are readily produced by the high pressure process. Useful films can be made from these materials at vinyl acetate (VA) concentrations up to 20%. Increasing the concentration of VA in the copolymer has the following effects:

(a) the level of crystallinity is reduced from about 60% at zero VA concentration to negligible crystallinity at a VA concentration of around 40%;
(b) the melting temperature and Vicat softening point are reduced (the Vicat falling to about 60°C at 20% VA);
(c) the modulus of the film is reduced;
(d) the toughness is increased, especially the low temperature toughness;
(e) the clarity is improved up to about 13% VA (though sometimes higher levels of antiblock additive mask this effect); and
(f) the compatibility with other materials increases.

These properties have given these films a market of some 195,000 tonnes per year in Western Europe. The majority (around 180,000 tonnes) is for copolymers with a VA content of 5% or less. These copolymers give films with properties not very different from LDPE.

At higher VA levels applications include additive bags containing preweighed quantities of additives (bag and contents being incorporated in a compound), bag-in-the-box and poultry packaging (use being made of flexibility and good puncture resistance), deep freeze bags and cling film.

One interesting use is for horticultural cloches and tunnel greenhouses. Ethylene/vinyl acetate film will allow sunlight to pass, but because of its lower transparency in the IR region, it will be more opaque to lower temperature radiation from inside the cloche or tunnel. It has been found that produce grown under EVA film can be marketed two weeks earlier than if grown under LDPE film.

6.4.2. *Ionomer Resins*

6.4.2.1. Structure and properties. Ionomer resins are so called because they contain ionic links between the molecular chains. The resin is made by high pressure copolymerisation of ethylene with a small proportion of a monomer which contains acid groups. The resultant copolymer chains are similar to LDPE but contain some carboxyl groups as side chains. The copolymer is blended with zinc or sodium salts to produce the ionomer.

It is thought that, in the solid state, effective crosslinks between copolymer chains are produced by small groups of metal ions and carboxyl units held together by ionic bonds. As the temperature is raised, these groups will disperse so that the ionomer can be processed as a thermoplastic material. Consequently, it is found that the viscosity of the ionomer melt shows a greater dependence on temperature than a typical LDPE polymer.

One advantage of the melt temperature/viscosity behaviour is that ionomer films have very good heat seal characteristics. Seals have a good hot strength (particularly of use in form, fill and seal applications). They can also be made in the presence of some oil or grease, especially if ionomers of higher ionic crosslinking are used.

Advantages of the solid film compared with LDPE are greater optical clarity and gloss (the ionomer has less crystallinity than LDPE), greater impact and puncture resistance especially at low temperatures, and good oil and grease resistance. The main disadvantage is its much greater cost compared with LDPE.

Ionomer films are used in packaging applications. Their oil and grease resistance make them useful for vacuum-packaging of meat for storage. They are also a very good blister and skin packing film. The ionic structure causes the resin to bond easily to other materials so that ionomer can be used as a bonding layer in blown film co-extrusion.

6.4.2.2. Processing. Ionomer resins will absorb water and so must not be exposed to the atmosphere before use. Equipment used for LDPE film is often suitable for ionomer film production provided motor power is sufficient. Brooks and Pirog[52] recommend an 18 kW drive for a 50 mm extruder and a 60 kW drive for a 90 mm extruder. Typical barrel temperature settings from feed to front zone are 150°C, 215°C, 205°C and 205°C. They also suggest a preferred screw design in which the feed and metering zones each comprise 25% of the screw length with a 50% transition section. Compression ratios between 3·5 and 4·0 to 1 are recommended.

Blow-up ratios similar to those found in LDPE film production are used with optimum film dart impact strength occuring at blow-up ratios between 2 and 3. The tougher ionomer film will increase wear on slitting knives. Care also needs to be taken with bubble collapsing and winding because ionomer films have inherently high friction coefficients and strong blocking characteristics. Slip additives are incorporated but take several days to migrate to the surface of the film.

6.5. Polypropylene (PP)

Isotactic PP produced by catalysed low pressure polymerisation is a semicrystalline polymer with a higher modulus and tensile strength than HDPE. When made by a conventional blown film process, PP film has a high level of crystallinity giving an opaque film. Consequently, most PP film is manufactured by the flat film process in which it is rapidly cooled from the melt to give a low level of crystallinity. Biaxial orientation of this flat film gives a clear, crisp packaging material.

Relatively clear film from the blown film method is obtainable if water is used to cool the bubble. In this method, PP is extruded at a melt temperature from 215 to 230°C to form a bubble moving vertically *downwards* from the die and cooled by water flowing over an annular weir. Excess water is wiped from the film by mechanical and air blades before winding. The need for a weir means that the process is less flexible in terms of film width than conventional air blowing methods.

Where clarity is not important the blown film process is satisfactory and is used to produce film which is subsequently stretched and split for weaving tapes, soft packaging twines and mechanically fibrillated yarns. In these applications blow up ratios are low (between 1·1 and 1·9:1) and the nip-roll to die distance is short to give bubble stability.[53]

6.6. Poly (vinyl chloride) (PVC)

6.6.1. Characteristics and Applications

Plasticised PVC of a thickness from 10 to 20 μm has several properties which enable it to be used as a distinctive packaging film. Firstly, it has outstanding optical transparancy and gloss due to its negligible crystallinity. Secondly, when plasticised (usually with DOA, i.e. di-(2-ethylhexyl) adipate) at levels up to 40 phr, the film is flexible. This enables it to be stretched by hand or machine to give a tight fit over an article to be packed. Thirdly, the combination of flexibility and low gauge gives this film a 'cling' effect so that it will adhere to itself or to other smooth, dust free surfaces. Also, plasticised PVC has elastic properties so that strained film will recover slowly when the deforming load is removed. This is useful in display packaging where finger impressions left in the film where the goods have been handled will recover to give a smooth surface.

Whilst rigid PVC provides a relatively good oxygen barrier, the addition of plasticisers increases the oxygen transmission rate. This enables plasticised PVC to be used for fresh meat packaging where oxygen is necessary to maintain a fresh, red colour.

Familiar applications for plasticised PVC film are for food packaging (often for stretch packaging of meat in expanded polystyrene trays) and for shrink packaging of greetings cards. Where shrink film is required, plasticiser levels are reduced.

6.6.2. Organoleptic and Toxicological Considerations

Since it is used for food packaging and because plasticised PVC is a compounded plastic, it is essential to ensure that no toxic ingredients can find their way to the customer. Poly (vinyl chloride) itself is non-toxic but it has been established that vinyl chloride monomer is carcinogenic. The level of vinyl chloride monomer is thus monitored in PVC film production. Nowadays, concentrations in incoming polymer are around 1 ppm with the level in rigid PVC products around $\frac{1}{2}$ ppm and less than this (probably as low as 0·1 ppm) in plasticised PVC. These levels are well below recommended and statutory concentrations which themselves are well below concentrations at which the toxic effects of vinyl chloride monomer have been demonstrated.

Lead, cadmium or barium stabilisers cannot be used in PVC film and the normal stabiliser system is based on calcium and zinc stearates. Epoxidised soya bean oil is also used as a stabiliser and plasticiser.

Migration of non-toxic substances from film to food must be considered so that packaging materials do not taint food products. Ingredients of plasticised PVC film are thus tested for odour before use. Contaminated plasticiser, for instance, may taint food wrapped in film containing the plasticiser. At present, proposed EEC legislation concerning global migration of substances from packaging materials is under consideration. In Germany the maximum plasticiser level is set at 22% in the film and maximum permitted concentrations in meat are 100 ppm. Manufacturers are readily able to comply with these limits.

6.6.3. Film Manufacture

Most producers of plasticised PVC blown film compound their own material using dry blend techniques. Single-screw extruders are used, with screws often containing mixing studs in the flight channel, similar to those used in PVC cable coating operations. The most common die used is bottom fed and designed to reduce material hold-up to a minimum. Despite this precaution, these dies have to be changed and cleaned frequently, because of a rapid build-up of additives and degraded PVC.

Bubble shapes are similar to those found in LDPE film production. Because of its 'cling' effect, handling of the film can be a problem and blocking may occur at the nip-rolls. Blocking is reduced by ensuring a low film temperature at the nip-roll and consequently the distance between nip-roll and die is kept large. Many applications require rather narrow film width and so in-line slitting is common.

For shrink applications, monoaxially orientated film is produced by a stretching unit positioned after the nip-rolls. The film passes over a heated roll and then, when at an appropriate temperature, is stretched by a faster rotating roll and nip which also cools the film. Biaxially oriented film may be produced by the double-bubble method. In this process the film is blown twice. The second blowing operation takes place after the first nip-rolls at a carefully controlled temperature which is lower than that of the melt leaving the die. Thus, the orientation imposed by the second blowing stage is more readily frozen in to give biaxial shrink film. Developments have taken place to control the first blowing operation in such a way that biaxially orientated film can be made without the necessity of the second stage.

6.7. Nylon Plastics (Polyamides)

Although expensive, nylon plastic films are very tough, provide a good gas barrier and have controlled clarity, varying from clear to semi-transparent. The relatively high melting points of these semicrystalline polymers (215°C for nylon 6, 250°C for nylon 66) means that they can be used at elevated temperatures, for example in oven cooking bags. Their main application is in packaging of foodstuffs and specialised products often in combination with another plastic material (see section 6.8). A large proportion of nylon film is made by the flat film process where it is often biaxially stretched (particularly in Japan). However, tubular blown film is also manufactured.

For processing, the nylon granules must be kept dry and not stored open to the atmosphere. Special extruder screw designs with a long feed and a relatively short compression section have been suggested for nylon but conventional designs are also generally satisfactory. Barrel temperatures suitable for nylon 6 would be in the range 220–250°C and at these temperatures a suitable motor power would be about 60 kW for a 90 mm extruder.

Side and bottom fed and spiral mandrel dies have been used successfully with nylon films. Normally, the die gap width is between 0·5 and 0·8 mm with a 30–50 mm land length. Care must be taken to ensure good melt temperature uniformity as the melt is of a low viscosity and hot spots will reduce melt viscosity drastically to give uneven stretching in the bubble.

Collapsing the tubular nylon film without creasing is difficult due to the relatively high stiffness and high friction coefficient of the film. Michael and Hort[54] discuss this problem and recommend the use of cooled collapsing frame surfaces which are made from perforated aluminium with a controlled degree of surface roughness. If processing conditions are modified to give a low level of crystallinity in the film (and hence good clarity and gloss), the coefficient of friction of the film will increase making the production of crease-free film more difficult. This can be a particularly difficult problem with nylon copolymers which have been developed to give high clarity by incorporating crystallisation-inhibiting units along their molecular chains. The high coefficient of friction can give rise to problems on subsequent processing machinery including the build-up of static electricity. Consequently, films are commonly passed under antistatic treaters and sometimes sprayed with fine powdered maize starch.

6.8. Combinations of Plastics

To optimise cost/performance characteristics, increasingly film containing more than one plastic material is manufactured. The plastic materials may be intimately mixed on a fine scale to produce a blend or kept as separate layers by a co-extrusion process.

6.8.1. Blends

There has been an increasing use of LDPE/HDPE blends to produce films with stiffness intermediate between the two types of polyethylene. More recently LLDPE polymer has been blended with the other materials. Blending is commenced by tumbling the different granules together and completed in the extruder much as masterbatches are incorporated. Sometimes a small proportion of EVA copolymer is also included.

Improvements in polyolefin film toughness (however, at the expense of clarity) have been made by adding styrene–butadiene block copolymers or EPDM rubbers to the polyolefin before film production. Table 5 gives results published by Phillips Petroleum Company showing how dart drop impact strength increases with increasing concentration of star-shaped styrene–butadiene block copolymers.

Recently, the addition of EPDM rubber to polypropylene has given

TABLE 5
EFFECT OF STYRENE-BUTADIENE BLOCK COPOLYMER ON DART DROP IMPACT STRENGTH (COURTESY, PHILLIPS PETROLEUM COMPANY)

Block copolymer concentration %	*Dart drop impact* (*g*)	
	160 μm LDPE film	*20 μm HDPE film*
0	249	110
5	322	192
10	511	216
15	767	227

films with improved impact resistance, particularly at low temperatures. There is also some evidence that such blends can show a greater resistance to tearing by raising the critical tear propagation speed above which the tear will propagate in a brittle rather than a blunt manner.

6.8.2. Co-extrusion

6.8.2.1. Applications. In co-extruded film, each layer retains the properties of its constituent plastic whilst the film as a whole combines the properties of the layers. Examples of blown film co-extrudates show how this combination of properties is advantageous.

Two different coloured layers of LDPE may be produced. For example, films with an outer white layer and an inner black layer have been used for packaging milk and for garden peat and compost. The outer white layer is used for display whilst the inner layer provides a UV light barrier for milk and reduces the dirty appearance of a compost bag.

A combination of HDPE with LDPE or EVA gives a strong, stiff film with easy heat seal characteristics on one side (or both sides if three-layer construction is used).

A combination of LDPE and nylon gives a film with good gas and water vapour barrier properties, heat sealable, with toughness and strength. Such films (typically 20 μm of nylon and 70 μm of polyethylene in thickness) are widely used for packaging cooked meats, cheese, chemicals and fresh meat (for long-term storage). They are usually converted to pouches or run on automatic vacuum-formed tray packing lines. Since nylon and polyethylene do not readily stick to each other, special steps have to be taken to promote adhesion between layers.

6.8.2.2. Co-extrusion plant. In blown film co-extrusion two (or more) extruders are used to feed a single blown film die in which separate channels in the die form concentric annular layers of the different plastics. The layers then join either just before, or just after, the die exit. Beyond the joining point the film passes through normal film blowing plant (apart from thickness monitoring) and behaves very much as if it were a single-layer film so far as its production is concerned.

Plant is available with spiral mandrel melt distribution for each layer and separate adjustments to obtain thickness uniformity around the bubble for each of two layers.

A blown film die fed by three extruders is capable of producing a five layer ABCBA structure in which, for instance, A is LDPE, B is an adhesion promoting plastic and C is nylon. Such a symmetrical layer configuration gives a film which will lie flat and not tend to curl up. Curling may occur in asymmetric constructions due to differential shrinkage or water absorption between layers.

6.8.2.3. Adhesion between layers. For most applications it is necessary for a good bond to exist between the layers of a co-extruded film. When layers are chemically similar, e.g. two colours of LDPE or LDPE/HDPE, the adhesive bond formed by the coming together of the two molten polymers is normally adequate. However, when two layers are chemically dissimilar (e.g. PE and nylon) then the natural bond is too weak. One of the earliest methods of promoting bonding between nylon and polyethylene was the Alkor process in which the two layers left the die in separated concentric rings joining together a short distance above the die. A reactive gas was introduced into the interlayer gap above the die face. It modified the hot polymer surfaces giving greater interlayer adhesion. A stronger bond is produced by having a third material (such as an ionomer resin) extruded as a separate middle layer between the nylon and polyethylene layers. The ionomer will bond to both nylon and polyethylene. In this case layers are joined before or at the die exit. Since the ionomer layer is not blown separately from the other materials it can be kept thin (say about 10 μm) to save cost.

6.8.2.4. Operation of co-extrusion processes. Temperature selection for operation when using layers of the same basic plastic is not difficult since it can be the same as for monolayer film production. For plastics which are normally processed at different melt temperatures, the die and one extruder have to be operated at the temperatures associated with the plastic requiring the highest melt temperature. Extruder temperatures for the other materials are normally selected to give a melt temperature higher than normal for these materials but somewhat below die temperature. Care must be taken in selecting a suitably high molecular weight polymer when operating at above-normal temperatures.

The total die output (the sum of the outputs from each extruder) is selected in the same manner as with a monolayer film line and control of film width, total thickness and thickness distribution follows monolayer procedures. The relative thickness of each of the layers is controlled by the relative extruder screw speeds and a knowledge of screw speed/output characteristics for each material. Whilst this gives the correct average layer thickness ratio, variations around the circumference of the bubble could give thin spots in any layer.

The monitoring of individual layer thickness distribution is possible. In one method, the co-extruded film is cut and observed edge-on in a microscope. This method depends on the boundary between layers being visible and is time-consuming. Where two materials do not adhere, the adhesive interlayer can be replaced by one of the outer constituents and the separate layers pulled apart and their individual thicknesses measured by dial gauge. This procedure can be useful during start-up. A more satisfactory on-line method for materials with differing IR spectra utilises an IR thickness gauge with two channels each tuned to a characteristic line in the spectrum for one of the layers. Though involving an initial capital outlay, this gives a continuous record of separate layer thicknesses. Pressure transducers fitted to each extruder between screen pack and die can provide a relatively cheap trouble shooting guide.

One of the main problems with co-extruded film is that it is often not possible to re-use film scrap and trim because the composite film is incompatible with every layer. In these circumstances, the reduction of unsaleable film to a minimum level has a strong influence on film cost. Minimising scrap can thus become a determining factor in the selection of operating conditions.

7. CONCLUSION

Polyethylene plastics are dominant in the blown film process and over recent years there has been a considerable amount of work on process modifications to improve plant output rates and film dimensional tolerances and to reduce scrap levels for these materials. This work has had success in improved die designs, more efficient cooling systems and the use of varying degrees of automatic control. The latter is the most recently introduced modification. It is likely to be developed and introduced more widely in the industry in the near future. Systems from relatively inexpensive width control to full microprocessor control of a number of film variables with central process data reporting facilities can be shown profitable and will become more widely used as converters are convinced of their advantages.

In the field of materials, the prospect for change at present seems to be for a growing utilisation of LLDPE. Also, the scope for co-extruded films and polymer blends is wide. One area may well be production of co-extruded films containing a very thin layer of plastic such as poly (vinyl alcohol) or a high nitrile resin which will provide excellent gas barrier film at a lower cost than present day composites.

From the point of view of research studies there has been some success in a continuum treatment of the film blowing process, correlating bubble shapes to polymer melt rheology. However, since most useful blown films are made from semicrystalline plastics, the development of solid morphology as the film crystallises from the strained melt should be important. Further investigation into this area together with a study of the morphology/film property relations should prove conducive to a greater understanding of the widespread but complex process of blown film production.

REFERENCES

1. SCHENKEL, G., *Kunststoffe*, **69**, 1979, 688.
2. ANON., *Eur. Plast. News*, February 1980.
3. GLASER, R., *Kunststoffe*, **70**, 1980, 600.
4. PASCHKE, E., *Kunststoffe*, **70**, 1980, 602.
5. ANON., *Technical Service Note A103*, ICI Ltd (UK), 1971.
6. DORY, R., *PRI Conf. Developments in Films for Packaging*, London, 1972.
7. FOUNTAS, N., *Plast. World*, May 1980, 64.
8. MARKS, C. J., *PRI Conf. Developments in Films for Packaging*, London, 1972.
9. ANON., *Leaflet 70–80/1e*, Alpine AG, Augsberg, W. Germany, 1979.
10. CATON, J. A., *Brit. Plast.*, **44**, April 1971, 140.
11. DAVIES, J. L. and GRAY, J., *PRI Conf. Developments in Films for Packaging*, London, 1972.
12. BARBOREK, F. J. and WILLIAMS, J. W., *Tappi Paper Synthetics Conf.*, 1973.
13. ZIELONKOWSKI, W., *Kunststoffe*, **57**, 1967, 607.
14. PLESSKE, P., *Kunststoffe*, **69**, 1979, 208.

15. ANON., *Web Guiding Book*, Mount Hope Machinery Co., Dartford, UK, 1970.
16. ANON., *Principles of the Mount Hope Roll*, Mount Hope Machinery Co., Massachusetts, USA, 1978.
17. FAINT, R. C., *Converter*, December 1973, 16.
18. WANZEK, R., *Kunststoffe*, **69**, 1979, 341.
19. HARLAND, W. G., *Plast. Polym.*, 1974, 162; *Polym. Eng. Sci.*, **7**, 1967, 58.
20. TORNBERG, I. E., *PRI Conf. Developments in Films for Packaging*, London, 1972.
21. ANON., *Travelling Flight Screw*, Sterling Extruder Corp., USA, 1981.
22. PROCTER, B., *SPE Journal*, **28**, 1972, 34.
23. KURZBUCH, W., *Plast. Eng.*, **30**, 1974, 43.
24. BRYDSON, J., *Flow Properties of Polymer Melts*, Iliffe, London, 1970.
25. CHENG, C. Y., *SPE Antec*, 1980, 57; CHENG, C. Y., *Polym.-Plast. Technol. Eng.*, **17**(1), 1981, 45.
26. AST, W., *Kunststoffe*, **63**, 1973, 427.
27. AST, W., *Kunststoffe*, **64**, 1974, 146.
28. FISCHER, E., *Kunststoffe*, **70**, 1980, 541.
29. HONSTRATER, R. A., *Plast. Technol.*, 1980, 99.
30. WAGNER, M. H., *Kunststoffe*, **68**, 1978, 362.
31. DOWD, L. E., *SPE Antec*, 1971, 274.
32. FARBER, R. and DEALY, J., *Polym. Eng. Sci.*, **14**(6), 1974, 435.
33. HAN, C. D. and PARK, J. Y., *J. Appl. Polym. Sci.*, **19**, 1975, 3257.
34. HOLMES, D. R. and PALMER, R. P., *J. Polym. Sci.*, **31**, 1958, 345.
35. NAGASAWA, T., MATSUMURA, T. and HOSHINO, S., *Appl. Polym. Symp. No. 20*, 1973, 275.
36. PETRIE, C., *Polym. Eng. Sci.*, **15**, 1975, 708.
37. WILLIAMSON, D. A., *Plast. Polym.*, **38**, 1970, 169.
38. CLEGG, P. L. and HUCK, N. D., *Plastics*, **26**, April 1961, 114; May 1961, 107.
39. KOHLER, F. J. and KRAUSE, J. M., *Kunststoffe*, **61**, 1971, 731.
40. PASCHKE, E., PLOENES, H. and DANZ, H., *Kunststoffe*, **67**, 1977, 377.
41. HOLLRICHER, O., LOCHNER, D. and ZIMMERMANN, H., *Kunststoffe*, **70**, 1980, 2.
42. HUGHES, M. J. and CAWKWELL, M. J., *Plastics*, **31**, 1966, 274.
43. RAKOVSKA, E., ONDREJKA, J. and KRKOSKA, M., *Plasty a Kaucuk*, **11**, 1974, 7.
44. PERRON, P. J. and LEDERMAN, P. B., *Polym. Eng. Sci.*, **12**, 1972, 340.
45. BAKERDJIAN, Z. and ILAHIE, M. Y., *SPE Antec*, 1980, 678.
46. JEFFRIES, M. P., *Brit. Plast. Rubber*, December 1981, 14.
47. EMMERICH, A., *Plast. Technol.*, **26**(2), 1980, 33.
48. SNELLER, J., *Mod. Plast. Int.*, **10**, July 1980, 30.
49. WRIGHT, W. D. and KNITTEL, R., *Tappi Paper Synthetics Conf.*, 1981.
50. ANON., *Polym. News*, **6**, 1980, 226.
51. O'NEILL, J., *PRI Symp. Developments in Polyolefin Film Technology*, Swindon, UK, 1981.
52. BROOKS, J. D. and PIROG, E. A., *Dupont Tech. Service Lett. STL70–4*, 1970.
53. ANON., *Technical Bulletin PP/30*, Shell Chemicals, 1973.
54. MICHAEL, D. and HORT, M., *Plastverarbeiter No. 10*, 1969.

Chapter 4

PVC WINDOW MANUFACTURE

R. G. BRUCE MITCHELL

Glass & Glazing Federation, London, UK

SUMMARY

This chapter is primarily concerned with the manufacture of windows with frames manufactured from hollow, unplasticised PVC profiles although a brief review of other, more traditional, systems is included. The particular requirements for PVC and its associated additives are discussed together with the relevant manufacturing technology. Because of the high quality and long life associated with uPVC windows, quality control testing has been discussed in some detail. Compounding and extrusion are then described and this chapter then concludes with a detailed discussion of the frame fabrication process.

1. INTRODUCTION

A well designed, well made and properly installed PVC window frame is an aesthetically pleasing, technologically efficient and cost effective building component. Window frames made from rigid unplasticised PVC have already achieved a proven record of performance in a demanding role, but despite this the technology is relatively new to the UK. PVC window manufacture involves the use of specially developed PVC resins, new compound formulations, sophisticated extruder designs, specialist calibration and 'Downstream' equipment, unique fabrication techniques and new test methods. In addition it has been necessary for new skills to emerge in designing PVC window frame profiles. The performance and service life of the finished product is dependent upon every stage of manufacture and this chapter follows the process through from the polymerisation stage to the installation of the window.

The following pages describe the materials and processes for the manufacture of window frames from hollow profiles. PVC is also used for sheathing timber profiles for window frames and in both solid and cellular form in combination with aluminium. Such processes of manufacture are however highly specialised techniques usually developed by individual companies and they are only briefly described in this chapter.

2. MARKET SIZE AND STRUCTURE

The size of the potential market for PVC windows is already encouraging manufacturers to invest in the not-inconsiderable costs of extrusion plant and dies for window frame profiles. Figure 1 is a forecast of the anticipated growth of the demand for PVC windows in comparison with other materials, and justifies the investment in manufacturing plant.

2.1. The UK Market

In 1980 the number of PVC window frame units installed has been estimated at nearly 3% of the total market. This will have required about 3000 tonnes of profile. If the market share reaches 20% then the demand for profile could reach 20 000 tonnes. In 1980 virtually every PVC window installed in the UK was wholly or partially made from imported components or raw materials.

The technology of PVC window manufacture, pioneered in Germany, has been well documented and the whole subject was reviewed at a conference held in Stuttgart in June 1979 entitled 'Twenty years of the PVC window'.[1] Many of the data in this chapter are therefore inevitably based on German expertise. German-made polymers and other additives are imported into the UK as well as the extruded profiles. Imported profiles are also brought in from Belgium, Denmark and Italy.

Overseas manufacturers operate in either of two ways in the UK. Some companies establish a sales office or agent to sell in this country.

Other companies license a UK company to market their complete system and possibly to extrude the profile in the UK.

2.2. Availability of British Designs

More recently a number of British companies have introduced systems specifically designed for the British market. Because every country has its own styles of architecture, window designs vary accordingly. Whilst windows designed overseas can be technically excellent and perfectly suitable for installing in British-built houses there are many people who prefer the traditional designs. For example most windows in Germany open inwards. In this country the tradition is for outward opening windows.

Traditionally designed British PVC window systems are now available but even so, much of the profile is made overseas to these designs. Companies who market a window system may or may not extrude the profile. They may or may not fabricate the frames themselves. The market structure is complex. To the end user two factors are important, namely the brand name of the system and the fabricator who not only makes the window but who installs it. From this it will be appreciated that the system designer from whom the brand name originates is a key operator. The system designer is the company who applies its expertise to designing the complete window system and its component parts, of which the profile is the most important. This company usually invests in capital goods for the extrusion of profiles—even if the actual extrusion of profile is subcontracted, money has to be spent on dies and die design. Money has also to be spent on extensive product testing and on advertising, sales promotion and a complete marketing operation.

2.3. Replacement Windows

In 1980 most PVC windows installed in the UK were replacements either in dwellings or in commercial and industrial buildings. There are several reasons for this. Many windows need to be replaced because the frames have either rotted or corroded. Property owners may want a replacement to avoid continual maintenance, to achieve a greater degree of comfort from a product made to a higher specification, to save heat losses through the use of a double glazed unit in a frame specially designed for this purpose, or to improve the appearance of the property. PVC is not the only material to give these advantages but will do so at a competitive price.

Prices paid for these high-performance windows are higher than those normally paid for windows in new buildings. New buildings are usually paid for with borrowed money whether in the private or public sector and maintenance, heating costs and other factors are usually someone else's responsibility. These are facts of life and the use of PVC windows in new buildings is likely to remain at a low level for many years to come. Even in West Germany after more than 20 years of successful use, 60% of the market for PVC windows is in replacements.

3. WINDOW FRAME PERFORMANCE

As a background to the technological developments that have occurred it is appropriate to summarise the basic requirements for windows. Those who pioneered the use of PVC for window frames had to market a product which met the existing standards of performance and durability. They did in fact introduce a product which was superior in many respects to established ones.

3.1. Types of Window

Types of windows in common use include fixed windows, or 'lights' as they are designated in the building industry, casement windows, sash windows and louvred windows.

Casement windows have opening lights which may be hinged at the side, at the top, or at the bottom: the building industry terminology is side, top or bottom hung. Some casement windows are hinged centrally so that they pivot either horizontally or vertically. A more recent innovation is the 'tilt and turn' casement window which can be opened in either of two directions with hinges incorporated on both the side and bottom for inward opening.

In sash windows the opening lights or sashes are designed to slide either vertically or horizontally.

The so-called 'french' window (more accurately defined by the building trade as a pair of 'door height windows') is hinged, but the modern counterpart which slides is usually referred to as a 'patio door'.

It is quite common for windows to be of composite construction comprising a combination of fixed and opening lights. For example, a window unit may comprise a fixed light with a top hung light or vent above it and hinged lights at one or both sides. The vertical members separating fixed and opening lights are known as mullions and the horizontal members between fixed and opening lights are known as transoms. The surrounding frame or frames may be of wood, metal or plastics. Before reviewing the properties of the materials it is however appropriate to summarise the performance requirements for windows of all types.

3.2. Performance Requirements

Windows have to be designed for a multiplicity of functions, which include admission of daylight, ventilation, protection against climatic extremes of heat and cold as well as against rain and air penetration, sound insulation and reasonable security against unlawful entry. In addition, opening lights must operate freely under all conditions and the window must be durable with minimum maintenance over long periods of time.

Specific requirements are detailed in the Agrément Board publication *MOAT No 1: 1974—Union Européenne pour l'Agrément Technique dans la Construction (UEAtc)—Directive for the Assessment of Windows.* These performance requirements are grouped under safety, habitability and durability headings as follows.

3.2.1. Safety

The integrity of the window must be maintained under loads imposed by wind pressure or suction, by vibration, by reactions from the bearing structure, by impact and by hygrothermal stresses. Windows must not present a fire risk. In addition, fixing devices and the opening mechanism must not impair safety under such loads. The window must also be designed so that cleaning can be carried out in safety.

3.2.2. Habitability

Air and water permeability is divided in three classes according to the rate of flow through the window at a range of wind pressures. Provision has to be made for water that penetrates the joints between sections to be discharged to the outside. Similarly provision must be made to discharge condensation water that may arise. For both purposes escape routes must be designed in such a way that water from outside cannot penetrate the inside under storm conditions.

Other requirements listed include provision for ventilation, sound insulation, appearance, lighting efficiency and ease of operation, including freedom from vibration caused through operation.

3.2.3. Durability

Requirements for safety and habitability must be maintained for the useful life of the window. *MOAT 1* also includes details for the determination of properties, which are necessary to establish that the window can meet the performance requirements. Tests include mechanical tests, permeability to air and water tightness, behaviour under variations of temperature and durability tests.

3.3. Materials of Construction

The basic requirements which have been summarised in the previous section must be met by all materials of construction. Figure 1 gives the market shares of windows with frames made from wood, aluminium, PVC and other materials. Some comments about these materials of construction are relevant to an understanding of the requirements for PVC.

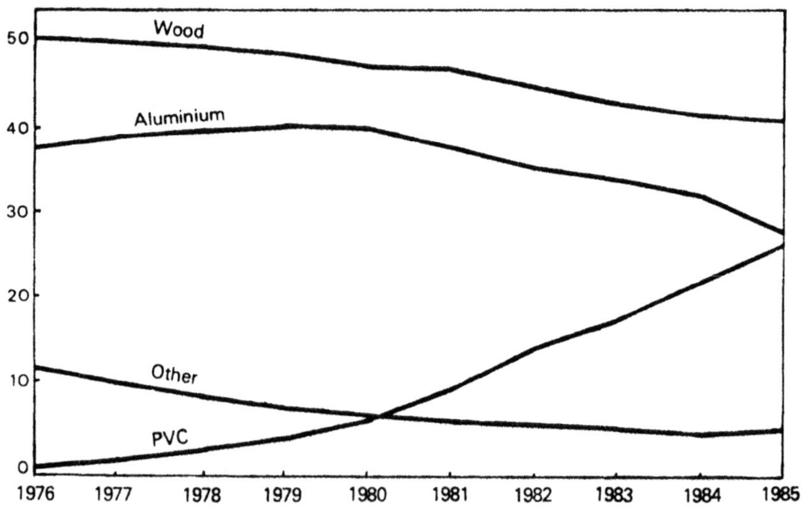

FIG. 1. PVC window frame market (as a percentage) (courtesy *British Plastics and Rubber*).

3.3.1. Timber Windows

The traditional wooden frame is constructed from softwood, mostly European redwood. Windows are available in a very wide range of standard and purpose-made designs and also in a wide range of dimensions. The ease with which timber can be worked means that it can be used for virtually any type of design without incurring a cost penalty. At the same time modern woodworking machinery and production methods can result in high output rates for standard designs. Softwood frames are usually supplied unpainted and unglazed to suit the traditional British technique of building around the frame before glazing and painting. Hardwood frames made from mahogany and other hardwoods imported from South East Asia and South America are beginning to take a share of the market, particularly for replacements.

A single glazed softwood window is low in initial cost but requires regular maintenance. A high-performance double glazed hardwood window has good appearance, but still needs limited maintenance and is in the same price bracket as a PVC window.

3.3.2. Aluminium Windows

The performance of an aluminium window frame, its design, its cost, the method by which it is made and the structure of the manufacturing industry are more closely related to the features of the PVC window frame than those of the timber frame. Aluminium window frames are fabricated from aluminium alloy extrusions by mechanical jointing, cleating or welding. Aluminium windows are generally fixed into prepared openings and can therefore be pre-finished. There are four main types of finish: mill finish (which is untreated natural aluminium), anodised finish (which is a uniform oxide coating), organic coatings (which are usually based on polyester or acrylic materials) and stainless steel finishes (which are light gauge stainless steel strips bonded to the exposed faces of the aluminium extrusions).

The advantages of aluminium frames are negligible maintenance and modern appearance. Fabrication costs are also less than for uPVC. Disadvantages are that they would normally have to be mounted in a timber subframe which needs maintenance, and aluminium windows without a thermal barrier may give rise to condensation.

3.3.3. Steel Windows

Steel window frames are manufactured from hot rolled sections mitred and welded at the corners. They can also be fabricated from hollow cold rolled profiles. They are usually protected from corrosion by hot dip galvanising. Thin gauge stainless steel strip may also be used for cladding mild steel. Steel windows, unlike aluminium windows, are normally supplied unglazed and they are painted on site, but like aluminium they are usually fitted into a timber subframe. Major improvements have been introduced in recent years through the use of plastics weatherstripping and factory applied finishes, but steel frames can also give rise to condensation.

3.3.4. *PVC Combined with Other Materials*

Window designs employing a combination of PVC (or other plastics) with wood or metals are very diverse. There are also designs employing combinations of two different plastics materials. Principal combinations can be summarised as follows.

3.3.4.1. Aluminium and PVC. In some designs, PVC profiles are secured to aluminium surfaces exposed to the inside of a building. The aluminium profiles have projecting fins which lock into grooves in the PVC profiles which are either a standard grade of solid PVC or alternatively a cellular PVC. The PVC profiles are mitred at the corners but they are not bonded together and they do not contribute to the strength of the joint. These designs are essentially aluminium windows where the PVC gives a decorative finish and helps to reduce condensation. In other designs developed for reducing condensation caused by the high thermal conductivity of aluminium, a so-called 'thermal barrier' is achieved by using twin aluminium profiles separated by a rubber or flexible thermoplastics gasket bonded to the aluminium.

Another design is based on the use of polyurethane foam injected into hollow aluminium profiles.

One combination of PVC and aluminium is used for an entirely different reason. For an improved colour range, coloured aluminium strip is used to clad PVC frames: in these designs the PVC is the structural component and the aluminium is there only to introduce a coloured surface—usually external to the building.

3.3.4.2. Other material combinations. Window frames have been made from flexible PVC extruded over steel and a highly successful design comprises a 1–1·5 mm PVC sheath extruded over a timber core (using a cross-head extruder and die).

Cellular PVC, polyurethane foam or even cellular concrete is used as a core material with PVC sheathing and some PVC profiles are themselves sheathed in a co-extrusion process with acrylic materials to give the wide colour range possible with acrylic.

Window frames have also been successfully made from GRP and from polypropylene oxide materials.

New designs are constantly evolving and they are intended mostly for specialist locations. Such designs also tend to be expensive and for these reasons the market share for most of these complex designs is likely to remain small. Combinations of materials with the best growth potential are PVC sheathed timber and PVC clad aluminium.

3.3.5. uPVC Windows

The frames of such windows are made from hollow rigid PVC profiles with one or more hollow chambers. Wall thickness is usually between 3 and 4 mm. Hollow metal channel tube or profile may be inserted in one of the hollow chambers to increase stiffness and minimise deflection under loads. Galvanised steel or aluminium is used for this purpose in frame profiles of 1 m or more in length. The metal does not form part of the corner joint although it may be screwed on to the PVC.

Design details vary and PVC window systems are available for every type of window that can be made in traditional materials. Most industrialised countries have evolved their own designs and in Germany there are over 60 PVC window systems on the market. The general principles of profile design and window frame construction are illustrated in Fig. 2.

4. SPECIFICATIONS FOR uPVC FRAMES

4.1. Approving Organisations

There are no national standards for PVC windows. In practice, however, a window system is not marketable unless it has some form of recognised approval. In Germany for example a window system needs to be approved by the Rosenheim Institute for Window Technology to their specification *RAL-RG 716/1—Quality and test specifications for plastics window frames and plastic windows.*

In addition it is necessary for the company which extrudes the profile to obtain approval by Suddeutsches Kunststoffe Zentrum (SKZ)—the officially approved test centre for plastics, for quality control.

Actual test procedures are carried out to DIN standards wherever possible but some tests such as corner weld strength tests have been specifically evolved for PVC profiles.

Approval to German specifications is accepted throughout Europe but a draft UEAtc Directive for the assessment of windows in PVC has

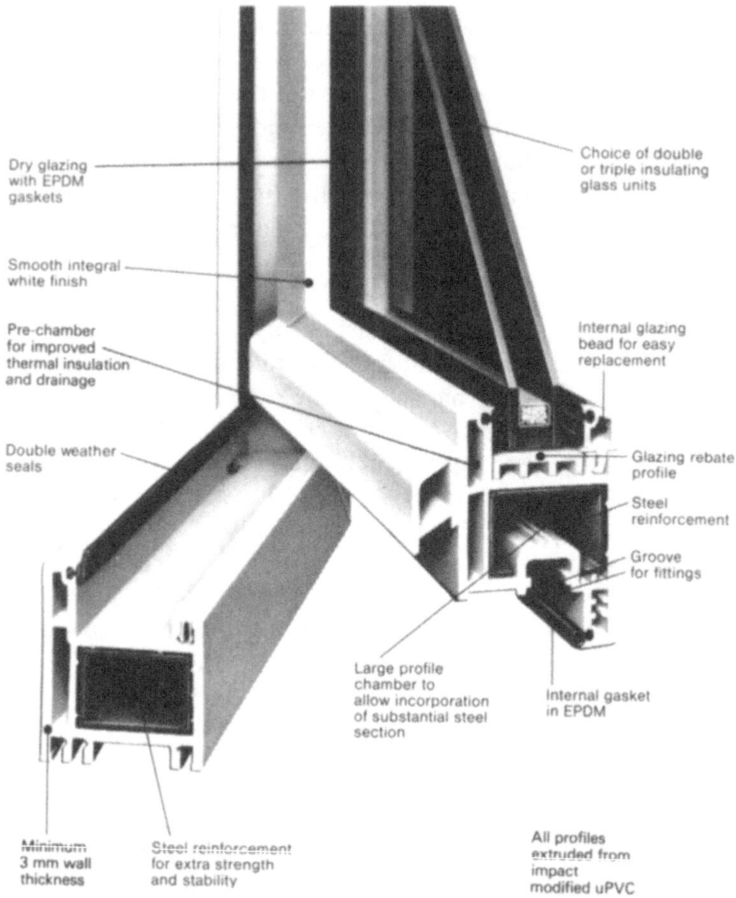

FIG. 2. Cross-section of a PVC window unit (courtesy Gebrüder Kömmerling Kunststoffwerke GmbH).

been prepared where the requirements are very close to those laid down in Germany. In the UK the Agrément Board has now published MOAT 17 based on the UEAtc Directive and a draft British Standard is being prepared. Test procedures in the UK will, where possible, be to ISO standards, or alternatively to British Standard specifications.

A new design will need an Agrément certificate for acceptance in the UK.

4.2. Summary of Specifications

The essential requirements of the proposed UK standard can be summarised as follows.

4.2.1. Scope

The standard colour white or near-white hollow profiles made from unplasticised PVC for use in the manufacture of window frames, door frames and patio door frames.

4.2.2. Raw Materials

Composition is defined as material made substantially from poly(vinyl chloride) with additives limited to those that are needed for satisfactory processing and achievement of required properties. Physical properties of the PVC material (from which the profiles are made) should conform to the proposed standards summarised in Table 1. Tests are to be carried out on pressed plaques prepared from milled sheet under specified conditions.

TABLE 1

PROPOSED STANDARDS FOR PVC-BASED MATERIALS

Test	*Test method*	*Minimum specified value*
Softening point	ISO R/306 (Vicat, 5 kg weight, 50°C/min)	75°C
Flexural modulus	ISO R/178 (rate 5 mm/min)	2 250 MPa
Impact strength	ISO R/179 (23°C)	20 kJ/m^2

In the extrusion of profiles designed for exposure to weather, the incorporation of rework will not be permitted, but for other profiles the principal requirement is that specification values for physical properties are met.

4.2.3. Profile properties

Standards for appearance and finish, manufacturer's identification, dimensional accuracy and a weight tolerance are detailed. Physical tests are to be carried out under the following headings.

(i) *Heat reversion.* Limits are given in order to establish minimum stress levels in the extruded profile. A value of 2% maximum is proposed for the test to be carried out at 100°C.

(ii) *Heat ageing stability.* This is designed to ensure the correct level of fusion in the finished profile. Performance is judged by ensuring that the profile shows no bubbles, cracks or delamination after ageing specimens in an oven at 150°C.

(iii) *Resistance to impact at low temperatures.* This is a falling weight test where impact resistance is determined by dropping a ball on to the profile which must not fracture (Fig. 3). It is likely that the weight will be a 1 kg ball with a 25 mm radius dropped from a height of 1 m on to a specimen 300 mm in length supported at 200 mm centres. The test temperature is usually standardised at values ranging from 0°C to −20°C.

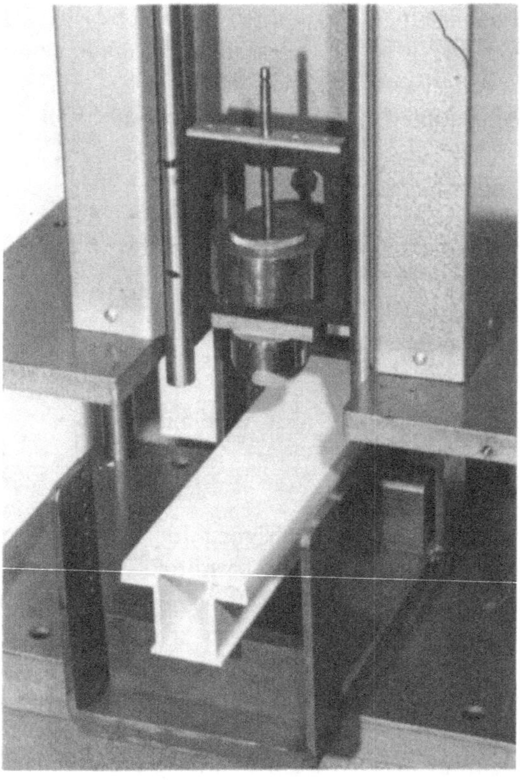

FIG. 3. Falling weight impact strength (courtesy British Industrial Plastics Ltd).

(iv) *Heat stability.* This test is intended to provide a rapid check that neither the formulation nor the processing conditions have reduced the effectiveness of the heat stabilisers. Reduced stabilisation will impair weather performance.

(v) *Weather resistance.* Specified accelerated weather testing conditions are to be laid down in order to determine colour fastness (using the grey scale) and the effect of weathering on impact properties (using the Charpy impact test method) where a reduction of more than 30% of the original impact strength would not be accepted. In the draft UEAtc Directive artificial ageing is determined in accordance with ISO/R 879 using a Xenon lamp: there are specified details of the lamp's output, weathering temperatures and humidity, wet and dry cycling and energy absorption of the specimen.

As few manufacturers have identical artificial weathering machines the energy absorption of the specimen is a particularly important factor in correlating performance. In specified conditions in the RAL standard satisfactory colour fastness and retention of impact strength can be obtained on specimens subjected to 6000 h in apparatus using a 6000 W xenon lamp. With the difficulty of standardisation and correlation of results of artificial or accelerated weathering, raw materials manufacturers would not normally introduce untried formulations without at least 5 years of natural weather exposure even if 5000 or 6000 h of accelerated weather testing had shown the material to be satisfactory.

(vi) *Weldability.* Weld strength is a measure of the suitability of the PVC material for welding. Test specimens are butt welded together on a hot plate welding machine and subjected to a standard tensile test. Performance is related to a weld factor, which is the ratio of the tensile strength of the welded specimen to that of a control sample. This factor is normally not less than 0·7.

Corner weld strength is a measure of the strength of the profile system and of the material. Nominal values will be supplied by the designer of the system and they will be determined from right-angle corner joints obtained by cutting samples of profile with 45° mitred joints welded together.

Alternative test methods can be used and will be decided upon by individual manufacturers. In one method the specimen is placed over a block of wood and pressure applied to the external corner of the weld, and in other methods the arms of the specimen are clamped in a tensometer and pulled apart (Fig. 4).

A reasonable joint will withstand a force of 750 kilopons but values as high as 1100 kilopons are possible. Even the lower value of 750 kilopons is higher than is obtained with other window frame materials.

FIG. 4. Weld strength testing (courtesy British Industrial Plastics Ltd).

4.3. Reasons for Tests

Tests described thus far are intended to establish that the correct materials are used and that adequate quality control is observed in processing and in manufacture of a welded frame. For full acceptance of a window system it must conform with standards for air and water tightness and perform the basic functions described earlier. In the UEAtc draft, standards are described for assembly, fixing of accessories (for example, pull-off strengths of hinges, pins and pivots), compatability with everything with which it is in contact, glazing procedures, installation and means of ensuring a seal between window and the surrounding aperture. In addition, limitations on use, such as maximum dimensions and service temperatures, must be declared by the manufacturer before supplying. Figure 5 illustrates the pull-out forces for different screw diameters in profiles extruded from compound based on Hostalit Z.

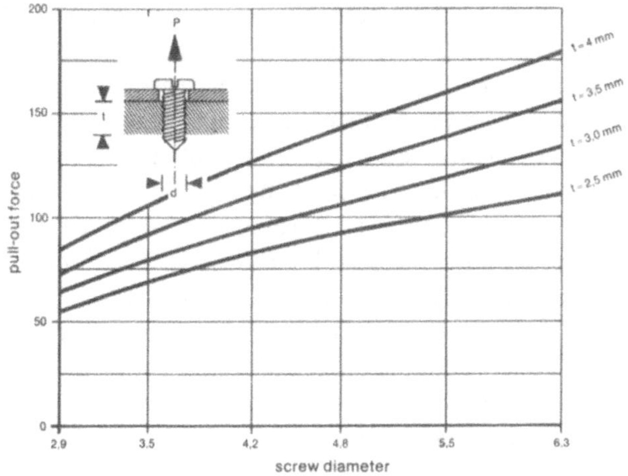

FIG. 5. Pull-out force for different screw diameters in Hostalit Z (courtesy Hoescht).

5. uPVC WINDOW MATERIALS

The performance required of a window frame clearly indicates the need for two vital properties. The PVC material from which the frames are manufactured must be tough and resilient and it must also be weather resistant with a long serviceable life requiring the minimum of maintenance.

5.1. Weathering of uPVC

uPVC has been successfully used externally for rainwater goods and other types of pipework for over 30 years. The good weathering behaviour is recognised, but it is necessary to state that an acceptable behaviour does not imply that no change takes place. All building materials, whether natural or man-made, will exhibit some change after exposure to the weather. Corrosion of metal, rotting of timber and fading and chalking of paint films are well known. Embrittlement of asbestos cement sheet on ageing and crumbling of fletton bricks from the action of frost are perhaps less well known outside the building industry.

Weathering is a complex process; apart from the damaging effects of frost, other important factors include elevated temperatures and temperature cycling, the UV light component and the angle of light incidence, rainfall, humidity and wind forces. The permutations and combinations of these factors are almost infinite and furthermore a building component that has acceptable performance in one climate may not be acceptable in another, because of the different effects from varying weather conditions. With PVC the principal effects of severe exposure are colour change, surface texture variation and embrittlement. These changes are greatly influenced by the choice of certain additives used in PVC compounds, and in particular by the choice of stabiliser and pigment systems. More detailed information is given on recommended stabiliser systems in later paragraphs. Extrusion conditions also have a marked effect on weathering performance of rigid PVC

Many uPVC formulations with acceptable weathering behaviour for pipework, and other building applications, such as cladding, do not possess the resilience and impact strength necessary for a window frame. uPVC is brittle at low temperatures and it is also notch sensitive.

5.2. Impact Modification

The search for acceptable impact modification has occupied PVC technologists for many years. The use of nitrile rubbers, ABS and MBS is well known and these modifiers are successful in a number of applications. For example, MBS is widely used in bottle compound formulations. Unfortunately, however, these types of impact modifiers cause a deterioration in the weathering performance of uPVC.

5.2.1. General Considerations

The first impact modifier for PVC to retain an acceptable weathering performance for window frames was chlorinated polyethylene (CPE). Because poly(vinyl chloride) and chlorinated polyethylene are closely related chemically they are readily compatible with one another. Hence the term 'polyblend' describes PVC polymers impact modified with CPE.[2] For window frame formulations polyblends containing 10% CPE are commonly used. Further work on impact modification has resulted in the introduction of ethylene/vinyl acetate (EVA) graft copolymers[3] and poly(acrylic ester) (PAE) graft copolymers. In addition acrylic modifiers have been introduced as additives for standard unmodified PVC polymers. Each system produces impact modified or impact resistant PVC (iPVC) with acceptable weathering properties but with minor differences in impact strength after exposure to various weather conditions.

The impact strength of uPVC is also related to the molecular weight: the higher the molecular weight, the higher the impact strength. Although high molecular weight polymers do not possess the resilience of impact modified or iPVC materials, they have the advantage of a high softening point. These polymers in the ISO viscosity range 115–130 are being successfully used for the manufacture of some window frame designs for warmer climates than in the UK, notably in the South of France.[4] These high molecular weight resins can still be modified if required with acrylic modifiers. The impact modified copolymer resins usually have ISO viscosities in the range 105–115.

Because of the importance of these various impact modifications they will be reviewed in more detail.

5.2.2. EVA Graft Copolymers

These graft copolymers are sometimes referred to as terpolymers because they comprise copolymers of ethylene and vinyl acetate grafted on to vinyl chloride. Ethylene and vinyl acetate copolymers are produced by radical polymerisation, using one of a number of processes. The vinyl acetate content is critical for achieving the required properties when grafted on to vinyl chloride. In practice optimum performance for window frame applications is obtained with an ethylene/vinyl acetate copolymer with 45% vinyl acetate content and high molecular weight (around 100 000), produced by a medium pressure solution process. It is a rubber-like granular material. The absence of carbon–carbon double bonds and sites sensitive to degradation by oxidation explain the good ageing and weathering performance when grafted on to vinyl chloride polymers by the suspension polymerisation process.

Resins are commercially available with EVA content as low as 5% or as high as 50%. For window frame formulations a 6% EVA content is most usual; this can be a resin produced with a 6% EVA content or a blend of a resin with a higher EVA content and a standard suspension resin. For example, a graft copolymer containing 50% EVA can be blended with a standard suspension resin in the ratio of 12 parts to 88 parts to give a net 6% EVA content in the blend. EVA graft copolymers are characterised as fine free flowing high bulk density powders.

With CPE and EVA modified PVC, investigations into the structure[5] have shown that the CPE or EVA envelops the PVC primary particles, so that a honeycomb type of structure is obtained, which can dissipate the forces produced through impact without cracking and results in a high notched impact strength. (If excessive shear occurs in processing a phase reversal can result and the modifier can form spherical or oblong particles in the PVC phase, which causes a degradation of notched impact strength.)

Some details of weather and impact resistance of EVA modified PVC have been given in a paper by Menzel.[6,7]

5.2.3. PAE Graft Copolymers

The poly(acrylate ester) may be an ethyl or a butyl acrylate, but with either type the process of grafting the acrylate ester on to the vinyl chloride during a suspension polymerisation process is similar to that used for EVA grafts. In the molecular structure PAE grafts are dispersed in the PVC as rubber-like spherical conglomerates which impart the required degree of resilience and impact resistance. PAE graft copolymers for window frame application normally contain about 10% polyacrylate ester. Blends of copolymers with standard suspension resins are possible to give intermediate impact strength properties.

5.3. Additives

5.3.1. Impact Modifiers

In addition to the use of impact modification at the polymerisation stage it is possible to add impact modifiers at the compounding stage. Certain types of acrylic impact modifiers impart the required degree of impact strength and weather performance in the extruded product. A typical acrylic polymer is a white fine free-flowing powder added to PVC polymers in the range 10–15 phr. An advantage of these impact modifiers is that they can be used with a wide range of standard suspension PVC polymers with ISO viscosity numbers in the range 110–130.

5.3.2. Stabiliser Systems

These may be based on lead, tin or barium and cadmium:[8] combined stabilisers are sometimes used, such as for example lead and barium/cadmium systems. With barium/cadmium (Ba/Cd) systems a secondary stabiliser is also used such as epoxidised soya bean oil (ESBO) and a chelator, usually an organic phosphite. In combination with barium/cadmium stabilisers ESBO improves light stability as well as heat stability and the chelator also improves initial colour and weather performance: another benefit is that it gives easier gelation.

The choice of stabiliser may depend upon a number of factors including efficiency for heat stabilisation, ease of processing, colour retention, weathering performance and retention of physical properties after exposure. Environmental and political considerations will also influence a decision as well as technical performance.

Barium/cadmium stabilised systems are widely used in Germany for window profile formulations, and tin stabilised compounds are preferred in the USA. In the UK lead stabilised compounds have always been widely used for rigid PVC, but despite their efficiency as stabilisers their use in window profile formulations is restricted to white compounds because they give rise to chalking after exposure to weather.

5.3.3. Lubricants

Choice of lubricants will be influenced by the stabiliser system used and will conform to usual rigid PVC formulation technology. Lubricants in common use include paraffin and polyethylene waxes, wax esters, stearic acid, hydroxystearic acid and calcium stearate. They must be balanced to give optimum internal and external lubrication—that is, optimum flow properties and lubrication of the compound against hot metal surfaces. Whilst the amounts of lubricants used are small their influence is considerable. Choice of lubricants and their balance is the key to commercial success of rigid PVC compounds.

5.3.4. Processing aids

Again these are used in accordance with the usual practice adopted for rigid PVC formulations. They give greater homogenisation of the melt at a faster rate which improves physical properties; they also eliminate 'plate-out' and reduce surface defects on the extrudate. The aids most commonly used are acrylic compounds.

5.3.5. Fillers

A small amount of calcium carbonate filler improves compound consistency and it helps to prevent the build-up of residues in the extruder. Normally the amount is less than 5 phr but larger amounts (between 5 and 10 phr) have been used with ultrafine particles of 0·8–0·9 μm; but this causes some reduction of elongation at break and a slightly adverse effect on the modulus of elasticity.

5.3.6. Pigments

Most formulations contain a rutile titanium dioxide which not only acts as a pigment but also improves weather stability by acting as a UV barrier. (Ultraviolet absorbers may also be included in formulations intended for use in certain climates and for coloured compounds.) The amount of titanium dioxide is usually in the range 3–10 phr. The actual proportion will be influenced by the choice of stabiliser and there are different opinions about the optimum amount necessary to give the best weathering performance.

There is a demand for coloured profiles and some manufacturers now offer a limited colour range, notably in brown and yellow. Organic pigments in concentrations around 1% are normally used. Coloured PVC window compounds do however introduce variables which are difficult to control. These variables include colour stability during processing, adverse effects on corner welding and reduced weather stability. In addition darker colours absorb more heat in sunlight, which increases movement as well as causing further effects on weathering performance. For this reason colour is sometimes introduced by means of a co-extruded poly(methyl methacrylate) (PMMA) sheath or skin or even by means of coloured anodised aluminium cladding applied to the external surface of the PVC profile.

5.4 Formulations

Because of the critical performance requirements and the inter-dependence of numerous ingredients (coupled with the influence of the extruder design), the formulation of a PVC window compound needs both expertise and experience. Raw materials suppliers will usually give detailed recommendations but as a guide the details in Table 2 indicate types of formulations which can be processed and made into windows with satisfactory performance.

TABLE 2
FORMULATIONS FOR PVC WINDOW COMPOUNDS

	Ingredient	*Concentration (phr)*
(1)	CPE/PVC polyblend	100·0
	Barium/cadmium (Ba/Cd) stabiliser	3·0
	Chelator (organic phosphite)	0·5
	Epoxidised soya bean oil (ESBO)	1·5
	12-Hydroxystearic acid	0·4
	Wax ester	0·5
	Polyethylene wax	0·05
	Process aid	0·5
	Titanium dioxide (TiO_2)	3·0
	Calcium carbonate ($CaCO_3$)	4·0
(2)	6% EVA modified PVC copolymer	100·0
	Ba/Cd stabiliser	2·5
	Chelator (organic phosphite)	0·5
	ESBO	1·5
	12-Hydroxystearic acid	0·4
	Wax ester	0·5
	Polyethylene wax	0·2
	TiO_2	5·0
	$CaCO_3$	3·0

TABLE 2—*contd.*

	Ingredient	Concentration (phr)
(3)	PAE modified PVC copolymer	100·0
	Ba/Cd stabiliser	2·5
	Chelator (organic phosphite)	0·3
	ESBO	1·0
	Polyethylene wax	0·4
	12-Hydroxystearic acid	0·4
	TiO_2	4·0
	$CaCO_3$	4·0
(4)	PVC suspension resin (ISO viscosity no. 110–120)	100·0
	Lead phosphite	4·0
	Calcium stearate	0·8
	Lead stearate	0·5
	Stearic acid	0·3
	Polyethylene wax	0·05
	Process aid	1·5
	Acrylic impact modifier	10·0
	TiO_2	4·0
	$CaCO_3$	4·0
(5)	PVC suspension resin (ISO viscosity no. 110–120)	100·0
	Tin maleate	2·5
	Calcium stearate	0·8
	Hydrocarbon wax	1·3
	Process aid	2·0
	Acrylic impact modifier	10·0
	TiO_2	5·0
	$CaCO_3$	2·0

6. COMPOUNDING AND EXTRUSION

Most window profiles are extruded from powder blends which are prepared on conventional high speed mixers.

6.1. High speed mixing

The ingredients are first blended in a heated mixer which brings the temperature up to about 120°C (the actual temperature depends on the formulation used). The mix is then transferred to a cooler mixer in order to reduce the temperature as quickly as possible (e.g. to 50°C or below). Some profile extruders prefer to use fully gelled or granular compound, which is necessary for extruding on a machine with a single screw. The granular material is cleaner to handle and extrudes faster than a powder blend, but the granular material costs more and because it has already been processed it does not retain the heat stability of the powder blend: this may affect weather performance.

6.2. Compound Properties

All the ingredients described influence the properties, and as already stated the choice of ingredients is a complex one. The different types of impact modification and stabiliser systems in particular affect the vital needs of high impact strength and good weathering. These ingredients also affect processing behaviour but because of the critical needs of the finished product the formulations arrived at over a period of 20 years or more result in very similar physical properties.

6.2.1. Typical Properties

A compound available in the UK based on an EVA impact modified PVC copolymer with a Ba/Cd stabiliser system and pigmented white has the typical properties summarised in Table 3.

TABLE 3
PROPERTIES OF TYPICAL COMPOUND BASED ON EVA IMPACT MODIFIED PVC COPOLYMER

Property	*Test method*	*Unit*	*Typical value*
Density	ISO R 1183	g/cm^3	1·40
Hardness, Shore D	ISO R 868	g/cm^3	78
Tensile stress at yield, 25 mm/min	ISO R 527	MN/m^2	49
Tensile strength at break, 25 mm/min	ISO R 527	MN/m^2	35
Elongation at break, 25 mm/min	ISO R 527	%	150
Impact strength (Charpy)			
23°C	ISO R 179	—	Not broken
0°C	ISO R 179	—	Not broken
Notched impact strength (Charpy), 0·25 mm notch			
23°C	ISO R 179	kJ/m^2	35
0°C	ISO R 179	kJ/m^2	8
Flexural modulus	ISO R 178	MPa	2500
Flexural yield strength	ISO R 178	MPa	79
Vicat softening point (5 kg load)	ISO R 306	°C	78
Coefficient of linear expansion	−20 to +60°C	cm/cm°C	8×10^{-5}
Thermal conductivity	Two-plate	W/mK	0·16

The properties and performance of the extruded compound are greatly influenced by the type and design of extruder used, by die design, by the calibration system and by the processing conditions employed. In addition profile design has a significant effect on window performance as well as imposing limitations on extruder die design.

An appreciation of the way in which processing conditions (such as mixing time and mixing temperature) influence impact strength is shown in Figs. 6, 7 and 8. These studies show that if an acrylic modifier is used, then the resultant compound is more tolerant to changes experienced during processing.

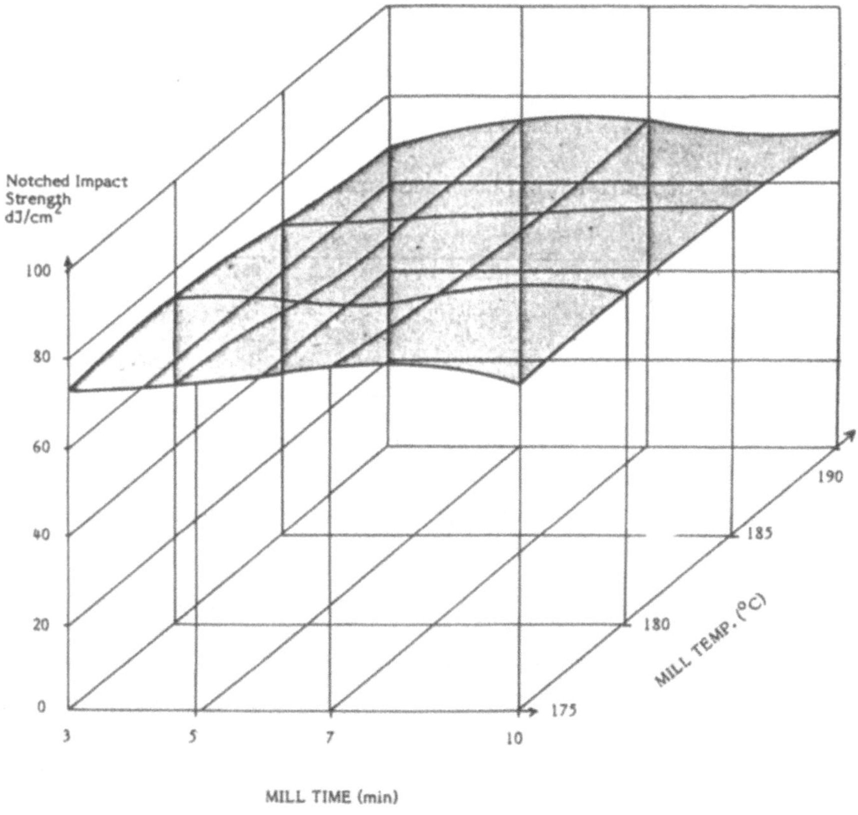

FIG. 6. Effect of processing on impact strength of compound containing 8 phr Paraloid KM 323B (courtesy Rohm & Haas).

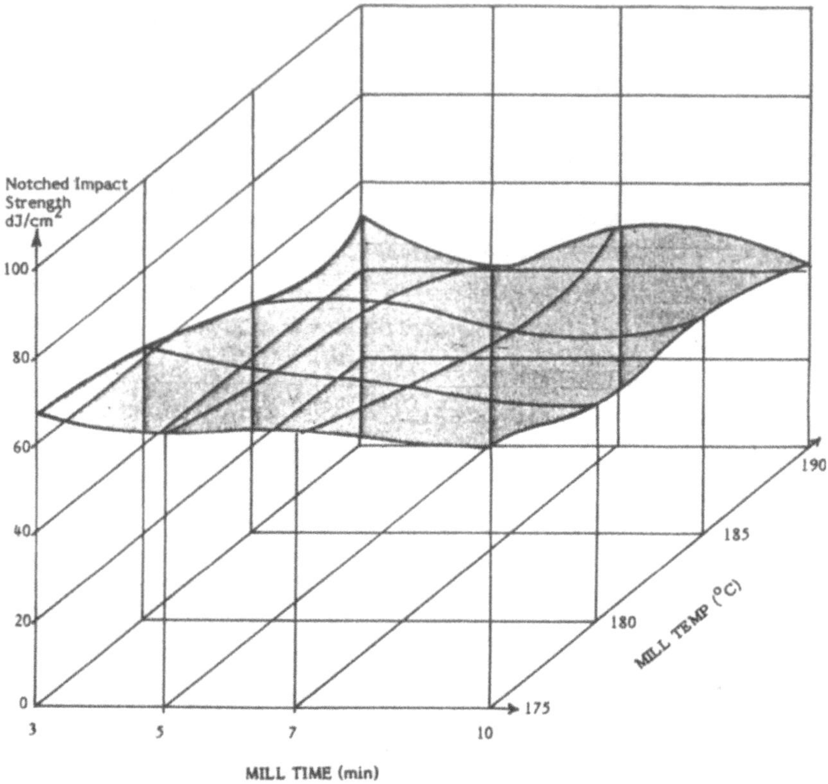

FIG. 7. Effect of processing on impact strength of compound containing EVA (courtesy Rohm & Haas).

6.2.2. Profile Design

Experience has created a number of design parameters which can be summarised as follows.

(i) In hollow profile designs the external wall thickness needs to be in the 3–4 mm range for satisfactory performance.
(ii) Thickness of external walls should remain constant.
(iii) The number of internal webs should be kept to a minimum.
(iv) Changes in section thickness must be minimised to avoid uneven flow in the extruder with resultant cooling problems and the risk of warping.
(v) External ribs should be as short as possible, again to avoid warping because any such ribs will be cooled on both sides.

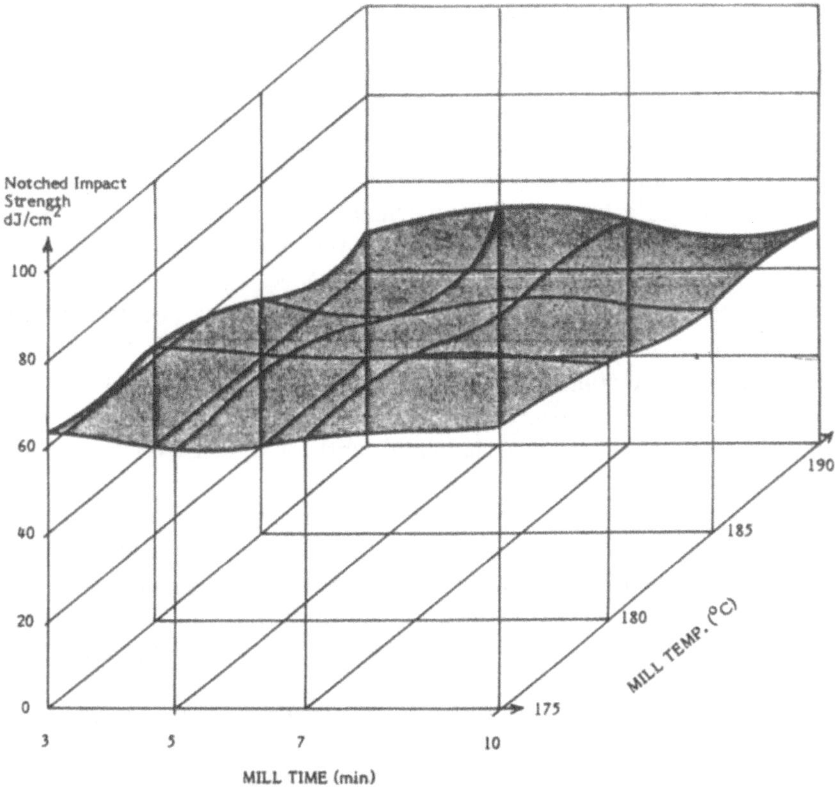

FIG. 8. Effect of processing on impact strength of compound containing 10% CPE (courtesy Rohm & Haas).

These parameters take into account the properties and processing behaviour of PVC and explain why a satisfactory PVC profile will not be a copy of designs in aluminium, in wood or in any other material.

Having made a choice of raw material and appropriate profile designs the choice of extruder and ancillary equipment must be made.

6.3. The Extrusion Process

The general principles for processing uPVC still apply to the extrusion of window profile compounds, but special factors to be considered are:[9,10]

(i) Window profile formulations are usually more sensitive to processing conditions than standard uPVC formulations for pipe and for other building profiles.

(ii) Profile complexity imposes special demands on die design.
(iii) Profile appearance and performance standards impose critical requirements for calibration and sizing.

6.3.1. Twin-Screw Extruders

Assuming the choice of a powder blend, a twin-screw extruder will be used. Most manufacturers of these machines have designed special screws for window profile compounds to ensure that good homogenising and mixing occurs and that a consistent melt reaches the die zone. Whilst many successful designs of extruder have parallel screws, an interesting development particularly suited to the extrusion of window frame components is the conical screw extruder. There are two types; in one, design screws have uniform depth of thread and in the other, a double conical screwtype, there is a changing depth of thread over the entire screw length.

Whichever design is selected it is necessary to recognise that the extrusion of window profiles with the required properties is likely to be at 70–80% of maximum extruder output. Profile must be produced at a rate which yields low heat reversion and which is resistant to warping during exposure to the weather. In addition, some compounds are sensitive to high shear in the extruder.

Another factor which influences output is cooling time for the complex profiles. If cooling is too rapid, strains will also be imposed.

In practice a maximum output rate for a main frame profile is likely to be about 1·3 m/min. Such a profile may weigh as much as 1·5 kg/m, which means that extruder output will be approximately 117 kg/h. Therefore extruders with screw diameters of 80–90 mm diameter will be needed for economical output of the largest section to be extruded.

The use of rework is not recommended in profiles exposed to the weather.

6.3.2. Die Design

This is a highly specialised skill. One machinery manufacturer has stated that his designers take into account no less than 30 values, but despite this no generally valid design data can be given for all profiles. A complicating factor is that in general tools have to be designed to suit the behaviour of specific grades of compound. The objectives of die design are however clear enough. They are to ensure that the profile emerges from the die at the highest output rate commensurate with even speed of flow across the section and to the correct dimensions. In practice dies are made from hardened chrome nickel steel which can be highly polished and which is resistant to abrasion and to corrosion (Fig. 9).

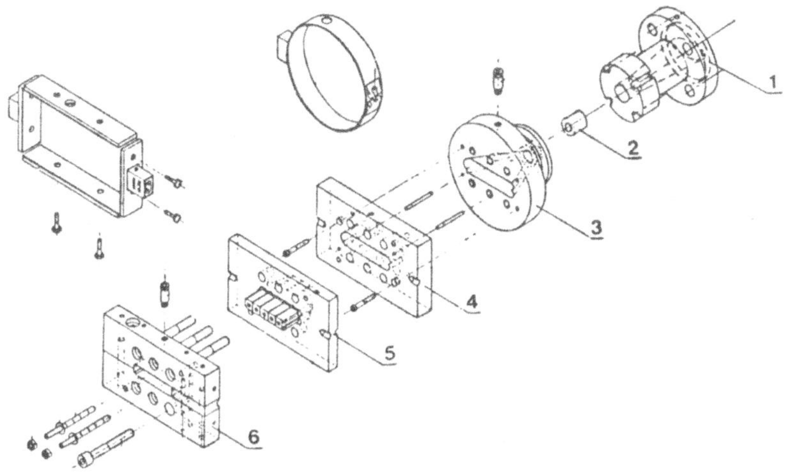

FIG. 9. Profile die (courtesy Krauss Maffei).

6.3.3. Calibration

Despite the need for satisfactory die design it is recognised that the calibration section of a window profile extrusion line is probably the most important part of the process. Design of the calibration section determines to a great extent the output and quality that can be obtained. Parameters which must be controlled are: the temperature gradient between the softened extrudate entering the calibration section and the hardened material emerging at the other end; the rate of cooling; heat transfer in the calibrating section, i.e. to the calibrating surfaces; and surface friction as a result of vacuum application. In addition the design must take into account profile construction and dimensions.

For hollow profiles dry calibration is usual and inevitably cooling takes longer than with an open profile. A total calibration length may be as long as 3 m and in practice individual sections of 200–700 mm are used. They are separated at distances of about 100 mm and water cooling may be applied in these regions. As many as eight calibration sections are known but three or four are more usual. Figure 10 is a schematic diagram of a calibration unit. It will be noted that the upper parts are hinged and can be locked for rapid setting. The function of the die unit is

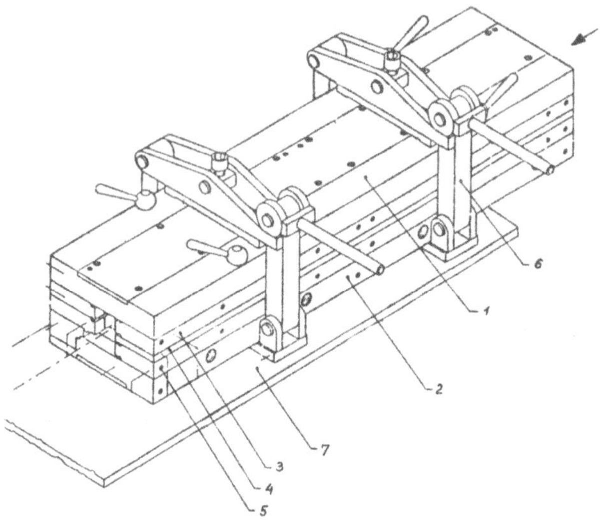

FIG. 10. Calibration unit (courtesy Krauss Maffei).

simultaneously to cool and shape the profile. It is therefore cored for water circulation and individual calibration parts have vacuum openings and slots.

Dies may be made from aluminium with an anodised surface, coated brass or stainless steel. Aluminium has the advantage of high thermal conductivity and ease of working but wear life is poor. Coated brass has better life and is probably the most widely used material, but stainless steel is the most durable.

6.3.4. Ancillary Equipment

The complete extrusion line will include a haul-off mechanism, usually of the caterpillar type. For large profiles the pulling force may have to be as high as 3 tonnes in order to yield satisfactory profile, but more recent designs of calibration units require forces only half this value.

The final operation on an integrated extrusion line is to saw the profiles to length and deposit the cut lengths, usually via a tipping chute, into a collecting rack. Saws are mounted in a travelling carriage so that the saw blade can move at right angles to the direction of extrusion and then cut the profile while it is held in a clamp.

6.3.5. Quality Control

Effective quality control procedures are essential to establish that the extrudate meets specification requirements. They are also a safeguard for the manufacturer by avoiding waste if control checks are made at regular intervals. Proposed UK standards, based on German practice, are as follows:

Test	*Frequency*
Appearance	Twice/shift
Identification	Twice/shift
Dimensions	Twice/shift
Low temperature impact strength (falling weight)	Twice/week
Heat ageing	Twice/week
Heat reversion	Twice/week
Corner weld strength	Once/week

In addition, longer-term control tests should be carried out after a specified artificial weathering period to determine colour fastness and the change in impact strength (using notched specimens for the Charpy method).

7. FRAME FABRICATION

The window manufacturer or fabricator will usually receive his PVC profiles in lengths in the range 5–6 m. After drawing from stores, the sequence of operations consists of sawing to length, pre-drilling and routing drainage slots according to window frame design, corner welding, finishing, fixing of metal attachments such as hinges, glazing and final assembly of component parts.

7.1. Output Rates

The time taken to produce an individual window unit varies considerably. As already stated, a window may be a simple fixed light or it may be a combination of a variety of fixed and opening lights. Output rates also vary for a given window combination according to the scale of operation. For example, a small fabricator with a crew of three men may produce twelve units a day on a profitable basis. One of the biggest manufacturers produces 2000 units a day. In all cases, however, the sequence of operations is similar, after transfer of the profile feed stock from stores. This sequence is described below.

7.2. Storage

Storage itself is important because storage conditions affect the performance of the finished window. Profile should be stored in an area adjacent to the fabrication shop and ideally at the same temperature. If this is not possible, one day's requirements should be kept in the fabrication shop for at least 24 h so that the profile is at working temperature. Profile should never be stored outside and it must be stored flat with solid support. Stacking height must not exceed 1 m so that distortion is avoided.

7.3. Sawing and Machining

As with storage conditions (and subsequent machining and fabrication operations) the saw cut influences the strength of the welded joint and long-term performance of the installed window.

7.3.1. Sawing

Specially designed saws are now available. Their diameter is normally between 300 and 450 mm and they rotate at a shaft speed designed to give a peripheral cutting blade speed of 50 m/s. For example, a typical saw with a diameter of 420 mm will rotate at 2275 rpm to give a cutting blade speed of 50 m/s. The saw blade is tungsten carbide tipped and the tooth form is specially designed for PVC with particular attention being given to the cutting angle. In sawing profiles to the correct length, double mitre saws are necessary to obtain the required degree of accuracy.

Optimum results are obtained with the use of computer controlled saws. They increase potential output as well as accuracy. Where V joints are needed, as they may be for welding mullions and transoms into the main frame members, combination saws are recommended so that the V joints can be cut in one stage.

7.3.2. Machining and Assembling

Before the lengths of profile are welded together, other machining operations can be more conveniently carried out at this stage. For example, slot milling and drilling of drainage holes and also holes for fixing lugs and hinges, locks, handles and other metal attachments may be performed at this stage. Specially designed machines are used for each of these operations including multi-head drills and milling cutters (as seen in Fig. 11, which shows how drainage holes are cut). Some manufacturers now fix mullions and transoms mechanically and instead of using a V joint they make a right-angle cut on the profile and then machine away the contour which matches the main frame member to which it is fixed. For this operation copy routers are used to cut away the relevant parts of the PVC profile.

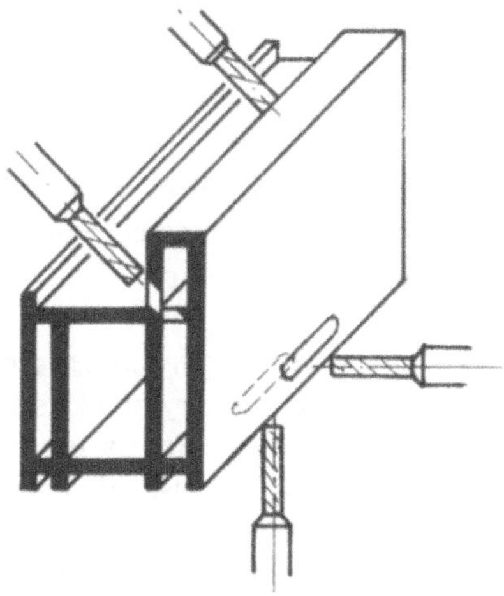

FIG. 11. Cutting drainage holes (courtesy Alu-Plast-Machine (UK) Ltd).

Where metal stiffening channel is to be inserted into one of the hollow chambers in the PVC profile, it is important that this operation is carried out at the correct stage so that the channel is in position for drilling if anything has to be fixed on to it. In most designs the metal channel is there solely to provide stiffening and does not contribute to the strength of the joints, but there are systems where the metal channel is screwed on the PVC and becomes a load-bearing reinforcement.

During sawing, drilling, milling, or any other machining operation, lubricants must not be used. This is because oil, grease or water residue might adversely affect the quality of the weld.

Weather stripping may also be inserted into grooves in the profile and can be welded together with the profile to give a good corner seal.

7.4. Joining of Profiles

Mechanical jointing, or screw fixing of joints, is also an established practice. It is particularly suited to the fixing of mullions and transoms and is often more convenient than the demanding conditions involved in hot plate welding a V joint. Screw fixing is also preferred in designs where the profile wall thicknesses are not sufficient to yield sufficient strength from hot plate welding. In practice, if the wall thickness of hollow profiles is less than 3 mm some form of mechanical jointing is necessary. However welding techniques are very widely used in the uPVC window frame industry. Both hot plate welding and solvent welding are employed although hot plate welding is the more important.

7.4.1. Hot plate welding

Hot plate welding of rigid PVC is not new although the process has few other applications. The essential feature of the process is to heat the sawn edges of the sections by softening them in contact with a heated metal plate and then to join them under pressure.

7.4.1.1. Welding machines. Welding machines are made for producing either one weld at a time or a multiplicity of welds simultaneously. There are machines that will produce as many as ten welds simultaneously. Machines are designed to produce right-angle corners and T joints but adjustments are usually possible for making joints through a range of angles. As well as providing a heated metal plate which can be raised or lowered into position, an essential feature of all machines is the clamping arrangement for holding the profiles and moving them into contact with the hot plate and subsequently into contact with each other. Many machines are equipped for computerised control of the weld cycle which, as in the computerised saw cutting operation, improves consistency of quality and output rates. Some welding machines are equipped with limiting devices for then trimming off the sprue formed during the welding process.

7.4.1.2. The welding operation. The cycle of operations can be summarised as follows.

(i) The prepared sections are clamped in the welding machine.
(ii) The hot plate is moved into position. The plate is electrically heated and is covered with a thin film of PTFE fo prevent the PVC from sticking to it.
(iii) The sawn profile edges are brought into contact with the hot plate until the PVC softens at the required temperature: see Fig. 12.

(iv) The profiles are moved away from the hot plate which is then retracted (or lowered, as in the machine illustrated).

(v) The softened profile edges are then brought into contact under pressure causing the PVC surfaces to fuse or weld together. During the welding action, 2·5 to 3 mm of profile is used up and forms the surrounding weld sprue. An allowance for this has to be made when determining the lengths of profile to be sawn.

(vi) Pressure is maintained for a cooling period.

(vii) In some fabrication processes the weld sprue is then trimmed by a cutting knife mounted on the welding machine.

FIG. 12. Hot plate welding of PVC profile (courtesy Plastal Ltd).

7.4.1.3. Weld strength and cycle times. Weld strength is influenced by the following factors: temperature of the hot plate; preheat time; profile temperature; fusion pressure; fusion time; cooling time.

As with solvent welding it is essential that the correct conditions are maintained if adequate weld strength (both initially and after exposure to the weather) is to be obtained. Weld cycles for profiles made from PVC would normally fall within the following ranges:

Hot plate temperature	210–250°C
Preheat time	30–35 s (including time of movement, which must not exceed 3 s)
Preheat pressure	1·5–2 kp/cm^2
Fusion pressure	4–9 kp/cm^2
Fusion time	35–60 s
Cooling time	35–60 s

Poor weld strength can arise from the following potential causes.

(i) The actual temperature of the hot plate is different from the indicated temperature.
(ii) The hot plate has been cooled down by draughts.
(iii) The cooling time is too short.

As with many plastics fabrication processes, an experienced operator can judge whether a satisfactory weld has been made by the behaviour of the PVC. In particular, the weld temperature can be judged by the appearance of the weld.

7.4.1.4. Finishing. Until comparatively recently, welded joints have been finished by grinding away the weld sprue and then buffing and polishing. The process is slow and therefore costly and today automatic knife trimming techniques are being used. These different techniques for achieving a smooth finish are described below.

The hand grinding and polishing technique is still extensively used for production of individual and short-run orders. Conventional disc grinders can be used, but with these it is preferable to grind in three separate operations using discs with different grit sizes in the ranges 120–150, 180–220 and 400–500 mesh. Alternatively, a specially designed hand sprue cutter can be used. With this tool, equipped with a frame which locates on the internal angle of the welded corner joint, a hard metal milling disc of adjustable height removes the weld sprue or seam. Again, two or three passes of the milling disc are recommended with the milling disc lowered after each pass until a clean finish is obtained.

After milling, a final disc sanding may still be necessary but the use of a sprue cutter initially is still quicker and more accurate than using disc sanders with coarse grit to remove the bulk of the weld sprue.

Finally, a high gloss finish is imparted through the use of a felt disc with a lambswool cover. For a satin finish a sisal polishing mop is used.

Hand finishing of welded PVC corner joints can take as long as

10–15 min for each joint and therefore considerable thought has gone into the development of techniques to reduce this time-consuming process without impairing the strength of the joint. It is important to realise that an incorrectly finished joint can reduce joint strength and can even lead to failure in service.

As a result of study, welding machines are now made with cutting knives which automatically trim off the sprue after the joint has been made. These knives will either trim flush or they can be designed to cut a small groove along the weld. This groove then becomes an intentional feature of the window frame in contrast with more usual practices designed to disguise the weld seam. Both techniques substantially reduce the total finishing time.

Weld sprue usually needs to be removed from other surfaces not exposed to view, as if it is left in place it may obstruct (a) the movement of an opening light, or (b) subsequent assembly and metal fixings. Such sprue is normally removed with a hand milling tool.

Another development is the introduction of finishing machines which carry out a complete finishing operation on each joint of a complete frame. In use the frame, supported on a metal jig, is fed into the machine which first simultaneously removes the weld sprues on the two faces with rotating milling tools. This operation is followed by three sanding discs passing over the weld seam to give the desired finish. Finally rotating cutters remove the weld sprues from the frame edges.

7.4.2. Solvent Welding

In this process right-angle injection moulded corner pieces are coated with a solvent adhesive and inserted into the hollow profile chambers. The process has the advantage of eliminating the cost of welding machinery. The initial joint is made rapidly but it must be left for 24 h to develop full strength before it can be handled. Final joint strength is less than that of a hot plate welded joint and in practice the solvent welding process is little used for window frame joints.

7.4.3. Mechanical Jointing

Joint detail will determine the appropriate location for screw fixing. With multi-chamber hollow profiles, screws can be driven into more than one web of the profile, but where this is not possible a metal component is inserted into the hollow chamber.

As previously stated, the cut end of a mullion or transom must first be shaped or contoured so that it seats on the outer frame. The location of screws will be arranged so that screw heads are hidden from view. Corrosion resistant self-tapping screws are employed. To simplify and speed up assembly time, there are available specially designed screws with a drill tip. This means that the screw drills its own hole and therefore pre-drilling is not necessary.

Mechanical jointing for right-angle corners is not the practice in Europe, but in the USA relatively lightweight profiles are used for vertical and horizontal sliding windows and in these designs a mechanically fixed butt joint construction is common practice.

7.5. Window Assembly

After completion of the jointing and finishing processes, the subsequent sequence of operations will depend upon the design of the window system and the method by which the frame will be secured into the window aperture.

As stated earlier, pre-drilling and slot cutting, e.g. for drainage holes, is best done before the cut profiles are joined together. After jointing and finishing it is logical to fit metal attachments wherever possible before glazing. Weather stripping may also be inserted before glazing. With hinges, handles, stays and locking devices part of the assembly must take place before glazing and part after glazing. In all cases the use of self-tapping screws with drill bits is a great advantage for metal attachments. The metal attachments must be designed so that the loads imposed upon them do not cause the screws to loosen.

Before glazing, provision can be made for securing the frame into the window aperture. For continued trouble-free performance, it is essential that the frame is secured to the brickwork or wall firmly and free of tension. In some instances the frame can be screwed directly into brickwork. In others, fixing lugs are used, where for example the frame may have to be placed over a wall cavity. The metal lug is secured to the wall and the frame then screwed on to the lug.

With either type of fixing the distance between fixing points should not exceed 600 mm and the distance from the corner should not be less than 150 mm. The window frame must be free to expand along the longitudinal axis of the profile. To assist in this screw holes should be drilled oversize or slightly elongated. The use of washers is recommended and screws should not be overtightened.

7.6. Glazing

Double glazed sealed units are normally used with PVC window frames but single glazing can be used if required. The sealed units usually comprise two sheets of 3 or 4 mm float glass separated by an aluminium channel which is sealed along the edges of the glass by a suitable sealant—either a polysulphide, epoxy polysulphide, polyurethane or a hot melt butyl. Dry air or nitrogen gas is used to flush out the air space before final sealing by means of a probe inserted into one of the corners. The edges of the units are usually finished with a tape. Thickness of the units is varied to suit the frame design but in practice air gaps between the inner surfaces of the panes of glass are generally in the range 6–24 mm. Some fabricators manufacture their own sealed units but they can be supplied by glass stockists.

Before fitting the glass it is necessary to prepare support packs for insertion in the glazing rebate in the extruded profile. These packs may be of rigid PVC, nylon, polystyrene or strips of laminated plastics, and are usually 80–100 mm in length and 2 mm wider than the sealed unit. They act as spacers and distribute the weight of the glass in the frame. They also prevent any direct contact between the glass and the window frame.

Dry glazing techniques are employed with synthetic rubber seals to hold the glass in place on both sides. Polychloroprene or ethylene propylene diene monomer (EPDM) rubbers are used.

The final operation is to snap into position the PVC glazing bead. This ensures a tight seal between the glass and the synthetic rubber strip. Contact pressure needs to be at least 0·8 kp/cm in order to achieve required weather tightness. The pressure, however, must not be excessive, and certainly not greater than 5 kp/cm or the sealed unit itself will fail.

Because some degree of water penetration can occur with dry glazed systems it is important that the PVC profile is designed to allow drainage and hence the need for drainage slots mentioned earlier.

More detailed information on recommendations for glazing are given in *British Standard 6262 Code of Practice for Glazing for Buildings* (formerly *CP152*).

7.7. Weatherseals or Weatherstripping

It is very important that the opening lights should seal effectively when closed against the supporting frame. To achieve this, similar sealing strips are inserted into grooves in the profile. Both edge seals and centre seals are used (Fig. 13). As already stated, EPDM weatherstripping can be placed in position before welding the profile joints. When this is done,

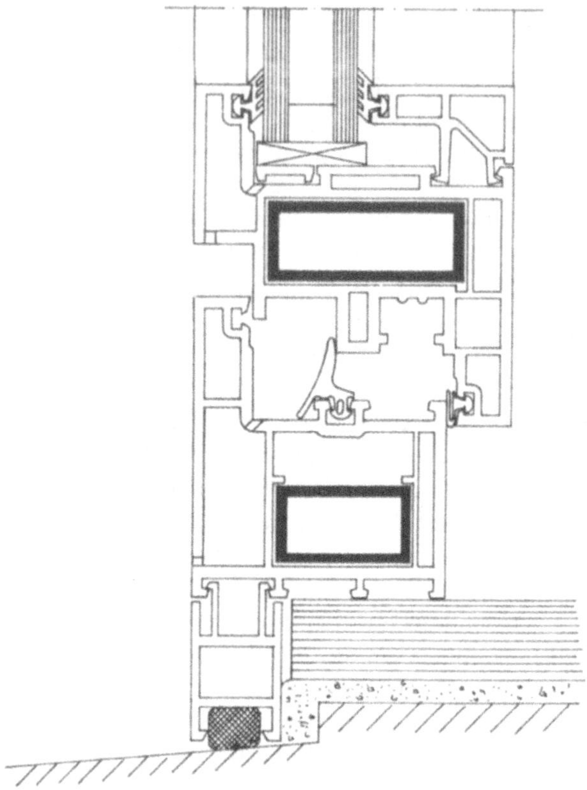

FIG. 13. Weatherstripping centre seals and edge seals.

the simultaneous welding of profile and weatherstripping ensures a good corner weather seal. The single central weatherstripping creates pressure equalisation and good protection against driving rain and it is therefore becoming more widely used.

When the glazed opening lights are located in the frame the last operation is to complete attachment of metal fixtures such as the hinges and stays. After this is done the window assembly is ready for delivery to the building site where it will be installed by specialist contractors.

7.8. Installation

The importance of the installation process must not be overlooked. Whether in a new building or a replacement in an existing structure the performance and service life of the window is very dependent upon this concluding operation. Window manufacturers will normally exercise rigid control and supervision of this. Failure is expensive and negates the technological effort behind over 20 years of PVC polymer development, formulation, extrusion and fabrication technology.

8. CONCLUSIONS

PVC windows are high technology products which can only become a commercial success because of the co-ordination of the design and manufacturing processes.

8.1. Technology

Required performance of the finished products has only become possible through the development of special polymers, additives and compound formulations. In processing, special consideration is necessary for extruder design, dies and calibration equipment. The fabrication process is demanding and the procedures must be strictly observed. Installation of the finished product requires high standards of workmanship and materials. Quality control during all the manufacturing operations must be rigorously carried out.

8.2. Co-ordinated Design and Manufacture

Whilst raw materials may be combined together on one site, the extrusion process carried out on another, and the manufacture of windows at a third location, choice of raw materials, machinery and design work must be co-ordinated. The integrated extrusion line demonstrates how a vital part of the process can be co-ordinated and a further step in this direction is the preparation of powder blends adjacent to the extrusion lines which some manufacturers are already putting into practice.

8.3. Marketing

A potentially worthwhile market has been identified but commercial success is dependent upon the factors already mentioned, together with investment in machinery and in promotional activity. There is enough increasing proof that with the observance of all these parameters a PVC window is a high performance product which matches the best of other types of window.

ACKNOWLEDGEMENTS

The author is indebted to a number of raw materials suppliers, machinery manufacturers, processors and window manufacturers who have helpfully supplied technical information, and in particular thanks are due to Mr R. S. Fitzpatrick, for his expert advice on the welding and fabrication of PVC windows.

REFERENCES

1. Conference Proceedings, *20 Jähre Kunststoffenster, Stuttgart, June 1979*, Institut für das Bauen mit Kunststoffen e.v., Darmstadt (German text).
2. FREY, H. H., translated from *Kunststoffe*, **67**(11), 1977, 706.
3. HARDT, D., *Brit. Polym. J.*, **1**, 1969, 225.
4. Conference Proceedings, *Journées d'Etudes sur la Profile Fenêtre PVC, Paris, October 1980*, Groupement de Promotion pour la Connaissance des Plastiques, Paris (French text).
5. MENZEL, G., HUNDERTMARK, G. and POLTE, A., 'EVA modified impact resistant PVC for window profiles', *Kunststoffe*, **67**(6), 1977, 339–40.
6. MENZEL, G., *Angew. Makromol. Chem.*, **47**, 1975, 181.
7. MENZEL, G., *Plastverarbeiter*, **24**, 1973, 397.
8. ANON., 'Stabilisation of window profiles', *Plast. Pub. Wkly.*, 16 August 1980, 8.
9. WESSEL, H., *8th Int. Extrusion Symp., Linz/Asten, September 1979*. Paper by Bernhard IDE, Ostfildern.
10. WESSEL, H., *8th Int. Extrusion Symp., Linz/Asten, September 1979*. Paper by Krauss Maffei.

Chapter 5

BLOW MOULDING PROCESSES

V. C. HIND, H. B. HALL and K. WHITEHEAD

Hamilton Machinery Sales Ltd, Gerrards Cross, Bucks, UK

SUMMARY

Blow moulding is a relatively young process which has developed steadily and now includes many hybrid processes which have in common only the single feature that they use a gas to inflate the soft plastic preform or parison. Machines using these variations on the basic theme are now available, in series production, and seem likely to establish themselves in specific market areas.

Major emphasis is placed on extrusion and extrusion accumulator processes as these represent the major part of everyday blow moulding.

1. INTRODUCTION

Continuous extrusion blow moulding and its development, continuous extrusion accumulator blow moulding, for large components, have been joined by a number of hybrid variations over the past several years.

One direct development is co-extrusion blow moulding, in which a normal one-material parison is replaced by a parison involving two, three or even five layers of different plastic materials, where the combined properties of the resulting laminate are significantly better or more economic than the original.

The advantages of true biaxial orientation are well known, but only over the past ten years or so have manufacturers offered equipment to produce bottles where the inherent properties of PVC, PET and PP can be more fully achieved by this process—more commonly known as 'stretch blow'.

Materials exhibiting minimal melt strength but nevertheless offering very desirable properties have led to the development of several types of injection blow moulding machine, where established injection moulding technology, including hot runner manifold systems, could be applied directly to the blow moulding process. A further variation on this method of obtaining a preform is via dip blow moulding, where a mandrel is inserted into a melt pool.

As the advantages of these developments are clear, it seems inevitable that further sophisticated combinations of processes will be developed, It can only be a matter of time before someone offers a machine producing blow moulded components via an injection, co-extrusion, stretch blow process!

2. MATERIALS USED IN BLOW MOULDING

In principle, any thermoplastic material which has sufficient melt strength to form a parison can be blow moulded. In practice, only very few common thermoplastic materials are used in significant quantities. Today, high density polyethylene and rigid poly(vinyl chloride) are still much the major part of total tonnage, although in the past few years other materials, for example poly(ethylene teraphthalate) for stretch blown bottles, have rapidly become significant. The more expensive engineering thermoplastics continue to find small specialised areas of blow moulding where their high price can be offset by the unique properties they offer.

2.1. Material Consumption Levels

The graph in Fig. 1 shows the growth patterns for all the major blow moulding plastics up to the late 1970s. It is clear that PVC and HDPE continue to find new markets, while PP has yet to achieve its predicted significance. Polyester (PET) is the only significant new growth material in the blow moulding sector, too new to appear on our graph.

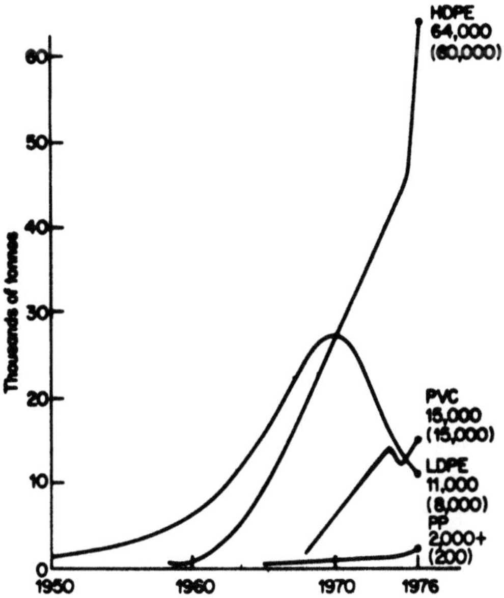

FIG. 1. UK blow-moulding market for polyolefins and PVC. The total tonnage for all blow-moulding activities is given at the end of each curve with the total for the production of bottles and containers given in parentheses.

2.2. Market Areas

In terms of the number of blow moulded units produced per annum, bottles are the most important—industrial containers, technical components and large blow mouldings complete the picture. Sailing surf boards of 4 m (13 ft) length are now in commercial production and tanks with capacities of over 10 000 litres (2 200 gal) have also been produced. Taking all types of bottles together, Table 1 sets out an industry estimate of end-use of principal groups in 1976. No more recent figures are yet obtainable. This suggests that 45% of total usage is accounted for by only three applications, cosmetics with toiletries, liquid detergent, and squash.

The reasons for selecting a particular thermoplastic for a specific end-use are normally a combination of price and correct properties. Table 2 shows in very simplified form the basic properties although we have deliberately avoided the thorny problem of price comparisons.

TABLE 1

ESTIMATED END USE OF PLASTIC BOTTLE CONSUMPTION 1976

End use	*Total number of units* (%)
Cosmetics/toiletries	18·5
Liquid detergent	15·0
Squash	11·5
Bleaches and lavatory cleansers	9·5
Shampoo	9·0
Household cleaners	5·0
Fabric softeners	4·5
Disinfectant/antiseptics	3·5
Motor oil	3·5
Cooking oil	3·0
Salt	1·5
Other food and drink	1·5
Other	14·0

Source: *Industry estimates*, British Plastics Federation.

TABLE 2

SIMPLIFIED TABLE SHOWING COMMON BLOW MOULDING THERMOPLASTICS PROPERTIES RELATIVE TO EACH OTHER[a]

Property	*LDPE*	*HDPE*	*PP*	*PVC*	*PS*	*PET*
Density (g/cm^3)	0·92	0·95	0·90	1·32	1·09	1·37
Chemical resistance	E	E	E	E	G	VG–E
Clarity	P	P	P–G	E	E	E
Toughness	E	E	G	P–G	P	E
Impact resistance	VG	VG	G	P–G	P	E[b]
Temperature resistance	P	G	G–VG	P	P	G
Moisture permeation resistance	G	VG	VG	G	P	G
Oxygen permeation resistance	P	P	P	VG	P	VG

[a] Abbreviations; E, excellent; VG, very good; G, good; P, poor.
[b] When stretch blown.

3. EXTRUSION BLOW MOULDING

Blow moulding processes which rely on an extrusion unit for the parison production stage still account for a great deal of blow moulding production.

3.1. Extruder Screws and Barrels

Extruder design has improved significantly in the past few years. The main aims have been to reduce melt temperatures and thus increase production speeds for the commonly used plastics, and to increase melt homogeneity or quality. Additionally, for a given diameter of extruder screw the maximum output available has increased.

Most extruders are of the single three-zone (feed, compression and metering) screw design with an L/D ratio of 20 or 24:1 with screw diameters from 30 to 160 mm. Most are powered by infinitely variable electric motors driven through V belt transmissions and two-speed reduction gear boxes (to provide different speed ranges for PE and PVC). Both ac and dc motors are in use. Some manufacturers also use hydraulic motor drives.

In most cases the extruder barrels have plain bores and are common to both PVC and polyolefin equipment, a change of extruder screw being the only physical alteration. For PVC, most manufacturers now supply 24 L/D ratio equipment, while for the polyolefins 20 L/D is still considered optimum.

Today's polyolefin extruders generally incorporate a grooved section in the rear of the barrel next to the feed zone, which is intensively cooled within precisely controlled temperature limits, typically at around 70°C. The screw used in this barrel is generally compressionless, with constant pitch and flight depth but equipped with special high shear sections, and quite unlike the three-zone screws employed with plain barrels (see Fig. 2). The usage of such screw and barrel combinations permits the production of melt with a low heat content (see Fig. 3).

In achieving 20 or 30°C reductions in melt temperature, the degree of mixing possible is severely limited. Masterbatch or pigment dispersion is thus more troublesome and can best be achieved by special screw designs which incorporate mixing sections along the screw length or at the front discharge end of the screw. A further advantage of the grooved barrel compressionless screw extruders is that their output is virtually unaffected by back pressure, and they can process any of the polyolefins efficiently, regardless of feedstock configuration (spheroids, cylinders,

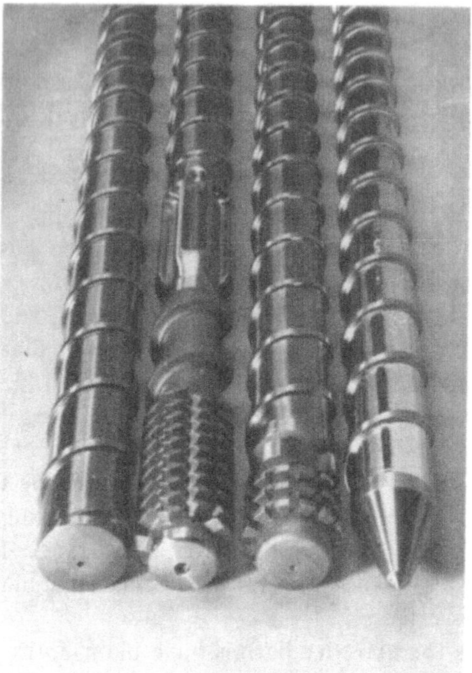

FIG. 2. Examples of modern extruder screws.

cubes, flakes, powder, or a mixture of any of these). To obtain the best advantage from such a system the mixing section must be adjusted to suit the job in hand. A typical system is shown in Fig. 4.

The exterior surfaces of the barrels of PVC extruders always incorporate helically wound copper cooling coils under the surface mounted electric heater bands. These coils are used for either air or heat transfer fluid circulation cooling. The cooling facility is actuated by three-point temperature controllers regulating the combined heating and cooling system via valves or contactors. Many manufacturers also offer individually mounted electric fan coolers for each of the three or four zones involved. Heat transfer oil circulation systems are recommended for PVC extruders exceeding 60 mm in diameter.

Extrusion screws are manufactured from high chromium steels, nitride hardened. Screws for PVC are additionally hard chromium plated on both flanks and roots. All the major manufacturers will offer non-standard extruder screws for specific applications (see Fig. 2).

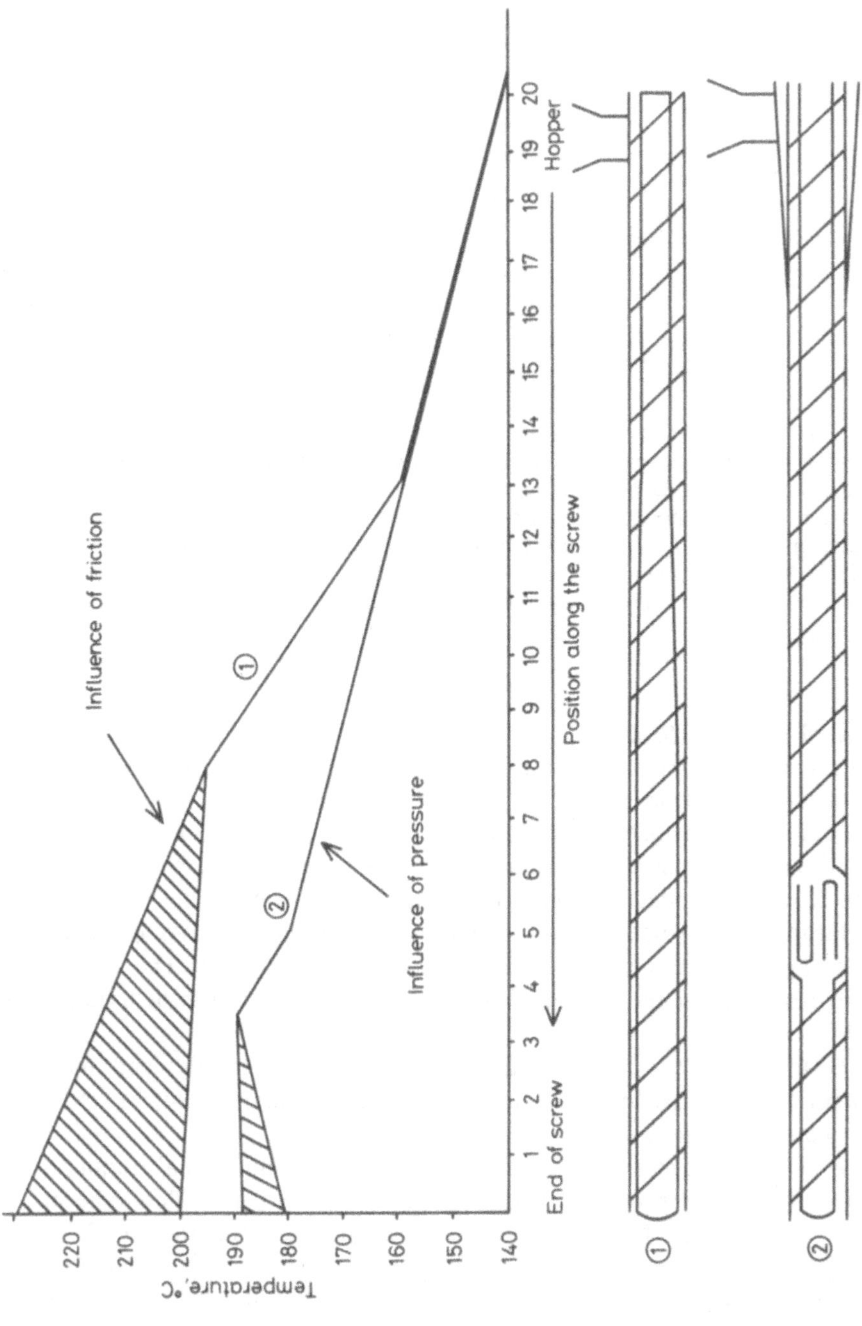

FIG. 3. Production of melt with a low heat content.

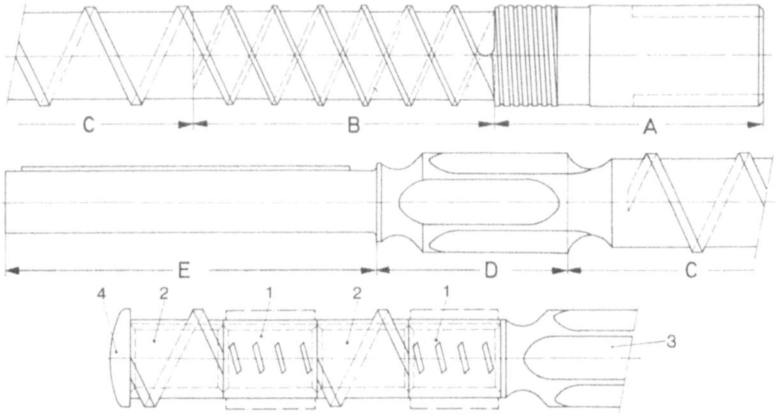

FIG. 4. Vario screw configuration showing one possible screw tip assembly. A, Shaft with labyrinth-packing; B, feeding area, double thread; C, compression area, single thread; D, blending part; E, Vario range = pump area (length *L*). The bottom diagram shows a standard line up of the pump area: 1. blending part, quadruple thread, 1/4 *L*; 2. conveying part, single thread, 1/4 *L*; 3. mixing section which fits onto plain shaft section E; 4, screw-cone. (Note: The optimal arrangement of the interchangeable parts for given production conditions can only be proved by trial.)

3.2. Extrusion Heads

The function of the head is to convert the plasticised material into a continuously extruded tube, or parison. There are different head designs for different polymers. Generally the simplest, and therefore cheapest, head designs are used for the more stable materials; more exotic designs are reserved for materials which genuinely need them.

Heads for polyolefins are of simple side fed design and have mandrels running through them from top to bottom (see Fig. 5). The molten polymer welds at the side opposite the in-feed. They cannot be used for PVC.

Heads for PVC are top fed, highly polished internally, and have torpedos supported by spiders (see Fig. 6). Flow of molten polymer must be laminar, without any stagnant areas.

A more recent move has been to universal heads on which all common polymers can be processed with mininal change of parts. This type of head is also preferable where parison programming is required for PVC. A typical design is shown in Fig. 7. Extrusion die and pinola or mandrel sizing and die design are important, and are based on the container being

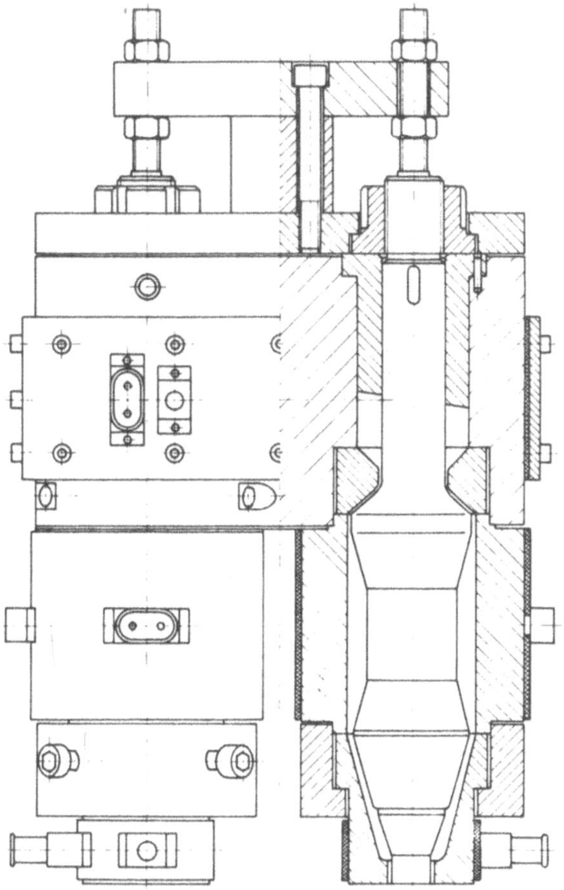

FIG. 5. Twin extrusion head for polyethylene.

produced. Oval dies or mandrels (incorporating both oval and round cross-sections) are now more frequently used and are employed in order to cope with the trend towards improved container quality at minimum weight. A high degree of experience and an empirical approach are needed to obtain the best results.

Another way of producing an oval shaped parison is to vary the die cross-section during the extrusion operation. One such system is the Moog PYDS.

3.3. Parison Cutting

The extruded parison has to be severed from the extrusion head once it

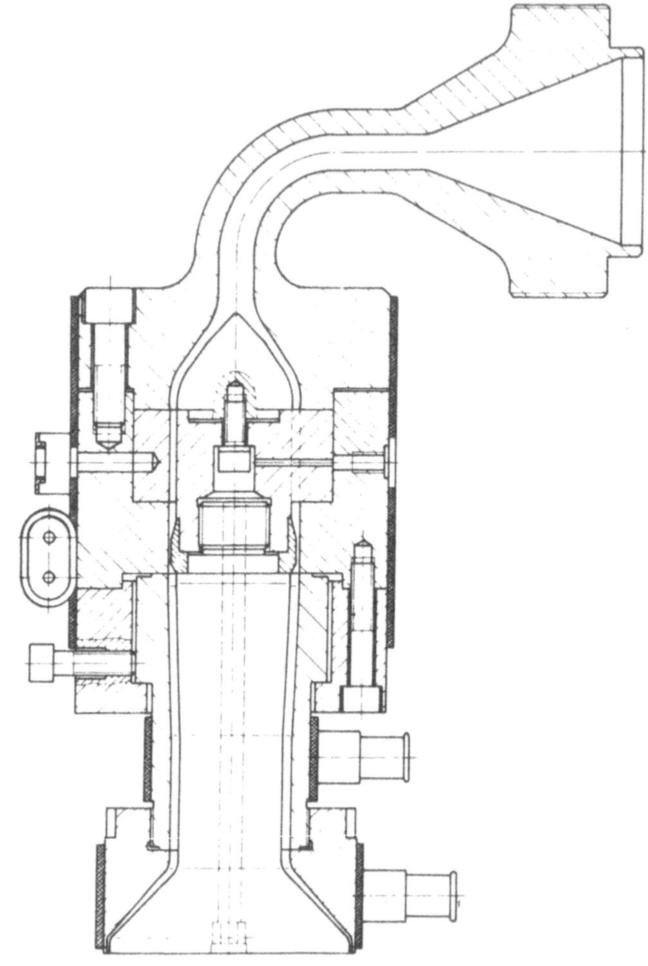

FIG. 6. Single extrusion head for PVC.

has been trapped by the closing mould halves. This can be achieved by one of the following devices.

(a) Hot blade or knife; generally a strip of special high resistance metal, clamped between two sets of jaws, and powered by a low voltage, high current transformer. Frequent setting and cleaning are involved. It is restricted in its use to the polyolefins and is

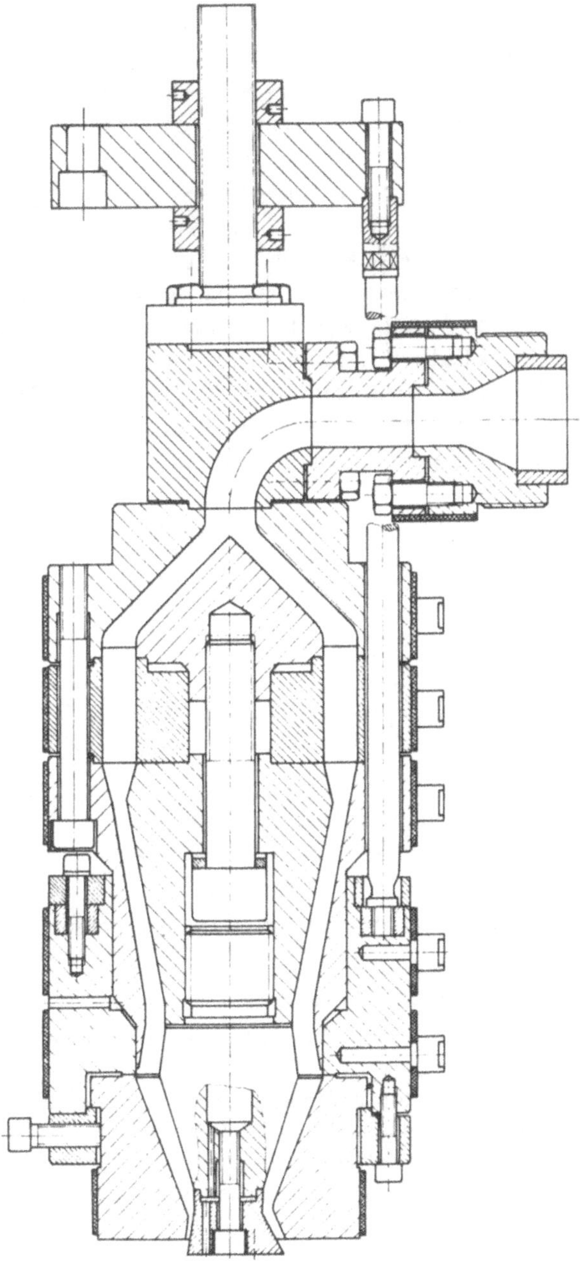

FIG. 7. Single universal extrusion head.

being increasingly displaced by the less troublesome cold alternatives.

(b) Cold cut blade (arrow head shaped). This is widely used for PVC and polyethylenes, especially when blowing the parison from within the neck.

(c) Combined cold cut with pre-squeeze. The leading edge of the parison is cut and sealed by means of horizontal blades and squeeze bars (see Fig. 8). This method is used mainly for high density polyethylene both where parisons fall outside the neck, and for handled containers. In conjunction with 'pre-blow' air, it allows controlled pre-inflation of the parison (improving circumferential material distribution) and enables the use of smaller diameter thicker parisons. It also reduces parison 'curtaining' problems associated with large thin parisons.

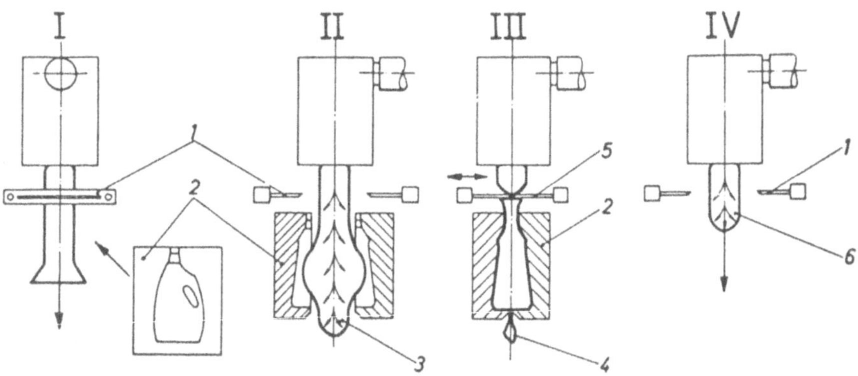

FIG. 8. Cold cutting assembly with pre-blow and pre-squeeze. Operating principle:

I (1) pinch cutters retracted, (2) open mould moves up,

II (3) parison has reached correct length, (2) the mould is fully up, begins to close,

III (2) mould is closed, (4) top and bottom parison ends pinched off by pre-squeeze device and mould, (5) pinch cutters advance and retract, and pinching jaws cut and weld the parison top end simultaneously,

IV (1) pinch cutters retracted, (6) sealed parison is extruded with controlled air flow.

3.4. Parison Control

Control equipment is used on general-purpose bottle blowing plant in order to give products which contain the minimum amount of material per bottle at high throughputs. Ideally the bottle weight should be maintained at just above the minimum specified.

3.4.1. Thickness

The demand for increased bottle quality at the lowest possible weight has led to the wider use of parison wall thickness control systems. Such systems are now becoming standard items on modern blow moulding plant. The original systems involving a very simple thin–thick–thin or *vice versa* change have largely been replaced by the much more sophisticated and versatile parison programmers. The majority of these involve three component parts: an electronic black box, a free-standing self-contained hydraulic power pack with sintered metal filter and a servo valve *cum* mechanical linkage to translate the electronic signals to the very small linear movements involved.

3.4.1.1. System Operation—Parison Programming. The system applies synchronised changes in the parison thickness over its length, to correspond with the varying degrees of stretch it undergoes during blowing, to maintain constant wall thickness in the finished component. During the extrusion of the parison, the wall thickness can be varied over its length at from three to 30 or more points (see Fig. 9). Most units are electronically controlled and incorporate a self-contained hydraulic power pack and servo valve system.

A typical system operates as follows. As soon as the initiation signal (normal blow mould close) is received, the distance to be moved by the servo valve, when changing the annular gap between the extrusion die and mandrel, is measured and the speed necessary for smooth movement from the first to the second step is calculated, based on a pre-set step period. The calculation is always made one program point ahead of that actually displayed.

The programmer thus calculates an ideal transition curve in advance. Modern programming units incorporate such features as display of program set points by light emitting diodes (LEDs), automatic cycle time synchronisation, digital selection of both time and measured die gap, visual indication of die position by LED band, warning LED if machine cycle is too slow, pushbutton for manual commencement of cycle and pushbutton for 'die gap maximum' for cleaning die.

3.4.1.2. Number of points. In general, as the number of programmable points increases (e.g. from 10 to 32), so does the ability of the unit to save on material costs.

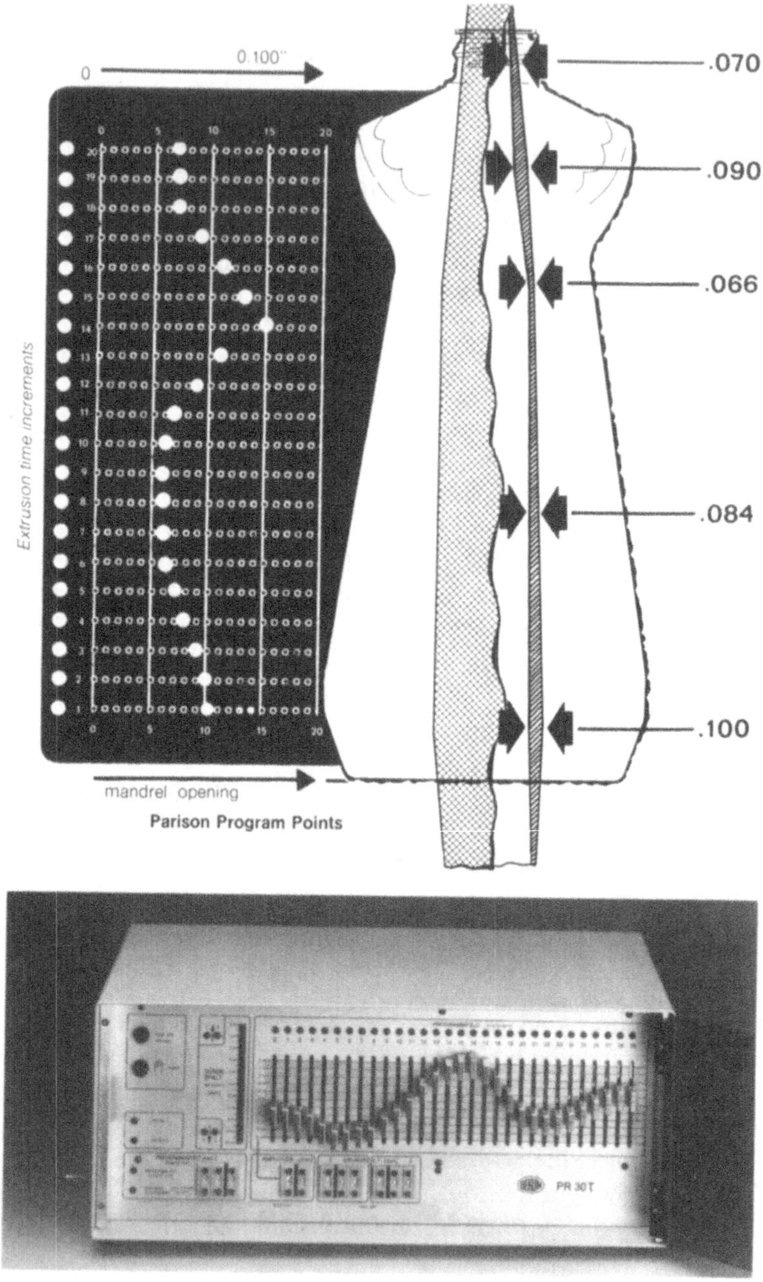

FIG. 9. Parison program control.

Probably the most important factor is the considerable saving of the material in the pinch-off area, where the drastic difference in the performance of the two programming systems becomes significant even for the simplest shapes.[1] Production results indicate that 2·5 g can be shaved off a 39 g high density polyethylene container at the pinch-off area when the programmer system is switched from a 20-point system to a 32-point system. The net material saving is 6·4%.

If other areas of the container are more precisely programmed, the top load strength of the container is improved.

3.4.2. Length Control

Such systems typically monitor, via a photo-cell arrangement, the length of the parison or parisons being extruded, store information and, when significant changes are detected over several cycles, automatically adjust the extruder screw speed to compensate. A length control system is therefore intended to minimise external effects, such as variations in material feed stock, external ambient temperature, mains electricity supply fluctuations and alterations by the operator to the parison wall thickness. It is generally used in conjunction with wall thickness programming.

As such systems are relatively cheap they will probably become widely adopted as they save on material. In bottle blowing they can save approximately 5% scrap material at the pinch-off.

3.5. Machine Movement/Construction

3.5.1. Machine Movement

Depending on the manufacturers' chosen system, machines are operated electromechanically or hydraulically. There is currently a move, following the established lead of injection moulding machine design, towards the use of proportional and digital hydraulics for faster and smoother machine movements. Improvements in machine speed have reduced dry cycle or 'dead time' by up to 50% (e.g. from 3 to 2 s), and this has significantly increased output; bearing in mind that most everyday plastics bottles can be made at between 6 and 13 s cycle times.

3.5.2. Mould Positioning

The mould platens carrying the single- or multi-cavity mould halves can be positioned relative to the extrusion head system in one of several ways.

(a) Mould movement is from the side or front (parallel with or at right angles to the extruder axis) either horizontally or diagonally or on an arc. This is a very popular system. Single- and twin-station units are available and can be used with single or multiple heads (up to four), fitted to single or twin extruders, e.g. Bekum, Kautex, Battenfield Fischer, Hesta, Hayssen, Magic, Morretti, Comec, Omea and Automa.
(b) The mould is on a rotary wheel or carousel system moving either vertically or horizontally. Machines using from two to 24 mould platens are available. The platens are fed by one or more extruders, e.g. Sidel and Graham.
(c) Moulds open and close under the heads, and the parison is accumulator extruded. Alternatively moulds open or platens are moved horizontally while the parison continuously extrudes, e.g. Uniloy and Bekum CS/CST.

3.5.3. Deflashing

Containers blown within the neck can be fully finished automatically in the mould, subject to the use of correctly constructed moulds with base deflashing and neck deflashing, or by the use of 'twist sleeve' blow pins. Handled offset neck or asymmetric containers in general must be transferred to a punch deflashing station. This can be within the machine frame and operated within the machine cycle or sited separately near the machine. The latter option involves the use of an orientated take-out conveying system for full automation.

3.6. Machine Control Systems

The gradual shift from electromechanical contactor systems to electronics is accelerating, and now incorporates microprocessor technology. All major machine builders offer memory programmable control systems. Bekum use the Phillips ISCOS extendable element system. Other systems have been developed in conjunction with Siemens and Barber Colman. They all allow for better efficiency, easier maintenance and trouble-shooting.

Microprocessor based sequence controls do not necessarily produce better containers. Nor do they speed up machine cycles. Their reliability is no better than that of solid-state controls.[1]

They have some basic advantages:

(i) Machinery builders can use the same controls on all of their machines, varying programs only. Manufacturing processes are simplified, volume is boosted, and costs are reduced. Microprocessor based programmable sequence controls should therefore ultimately reduce the cost of machine control to the user.

(ii) Microprocessor based sequence controls can troubleshoot themselves and machine components such as limit switches and solenoids.

(iii) Safety features, including those specified by customers, can be added at low cost. Industry and government standards can be readily implemented.

(iv) The memory of most microprocessor based sequence controls is large enough to accept the sequencing of auxiliary equipment. Robots, trimmers and material feeders, for example, can be brought into the machine cycle, making downstream equipment synchronous with the machine and less costly to control.

(v) Reprogramming machine sequences to suit special conditions is easier than it would be with solid state controls. Machine sequence can be altered by a field programming device to incorporate special machine movements or sequences. Users can control programmes to adapt to modifications and even entire new processes. Some controls load machine set-point, process set-point and sequences into sequence controllers by magnetic cards, tapes or preprogrammed chips representing the set-points used in previous runs.

3.7. Calibration

Most machines use the top blow calibration system. Increasing demands for high quality neck finish and reduced maintenance costs (neck striker plates, cutting bushes), and reduced downtime has led to the adoption of hydraulically operated blow pin movement for neck calibration. By the further sophistication of three-step calibration, neck flash can be completely separated and the point in the cycle at which this is ejected can be regulated so that the chance of scarring a finished container as it contacts hot sticky flash is eliminated.

3.8. Moulds

3.8.1. Mould Cooling

Work continues on increasing container output by improving mould cooling, as ultimately it is the time taken to cool the inflated parison in the mould which governs output rate. Modern moulds have cooling channels as opposed to flood cooling. These channels are evenly spaced around the cavity to ensure uniform cooling over the whole surface. Channel size depends on the size of mould and is typically 6 to 10 mm in diameter. A high volume flow and low resistance to water flow through the mould are essential (see Section 3.8.4).

3.8.2. Materials of Construction

Many different materials have been and are used for blow moulds, including hard or aircraft aluminium, chromium steel (for PVC), stainless steel, beryllium/copper (Belan) and Kirksite. For longevity and durability we recommend chromium steel for PVC and stainless steel for polyethylene, with possibly beryllium/copper (Be/Cu) inserts in the neck and base areas, depending on the container in question. Aluminium, although theoretically worthwhile compared with steel because of its thermal conductivity and weight advantages, is less popular than might be expected because of the disadvantage of relative fragility and difficulty of repair.

3.8.3. Base and Handle Design

Base and handle pinch-off design, and the compression area behind the pinch-off blade, are of critical importance, and the design used depends on the article size, the polymer selected and the machine on which it is to be run. Most machines allow adjustment of mould closing speed, especially over the last few millimetres. If the flash is not compressed correctly it leads to weak pinch-off areas or very hot flash, which can subsequently stick to the container after ejection. Base deflashing can best be achieved by fitting base 'flipper' deflashing units which operate before mould opening and therefore separate flash and container completely.

For handled containers where flash sometimes accounts for up to 60% of parison weight, some detailed improvements have recently been developed which are essential when running fully automatically using punch deflashing systems. The top and base flash areas can both be ribbed to improve cooling and to give rigidity to the flash. The 'eye' of the handle must also be adequately cooled over its whole area, which means it must be compressed slightly to be in contact with the mould surface.

A common problem with handled containers is webbing in the handle area. This manifests itself as an unwanted membrane partially blocking the handle internal area. It can always be overcome by applying correct design parameters to the relationship between handle width and depth, a subject outside the scope of this chapter.

3.8.4. Blow Pin Cooling

Blow pin design and cooling are important, as cooling of the neck section (the thickest part of the container) is a limiting factor on output rate. For highest efficiency, water cooled blow pins are therefore, essential. It is now possible to cool blow pins as small as 6–8 mm diameter down to their tips. Adequate blow pin cooling also reduces neck 'push-down' internally. The introduction of a refined design of detachable-tip blow pin has reduced downtime when replacing tips and/or cutting bushes. These can be removed and replaced easily without affecting the established machine calibration position. The outer sleeve on the blow pins and/or the outer edge of the cutting bush can be undercut to allow use of automatic neck deflashing systems (see Fig. 10).

Refrigerated water systems coupled to the blow moulds are essential to ensure optimum output. Each machine ideally should have its own unit for both mould and blow pin cooling to achieve and maintain the correct temperature for the particular mould/polymer combination. The mould temperature should be maintained just above the dew point to avoid condensation on the mould surface and subsequent marking of the finished components. Typically PVC moulds should be run at 4–6°C, and HDPE at 6–10°C.

3.8.5. Internal Cooling

Output increases of up to 30% have been claimed due to internal cooling of the container during the blowing cycle. Intensified internal cooling may be achieved in several ways.

3.8.5.1. Carbon dioxide (CO_2) *cooling.* This is the best known method at present. The parison is blow moulded to the shape of the cavity with ordinary air and then liquid CO_2 is injected into the container where it evaporates and picks up heat. Its profitability is in dispute as imprecise dosage of cooling liquid results in excessive consumption: the same comment applies to N_2 cooling.

3.8.5.2. Nitrogen (N_2) *cooling.* This is a relatively new method. The

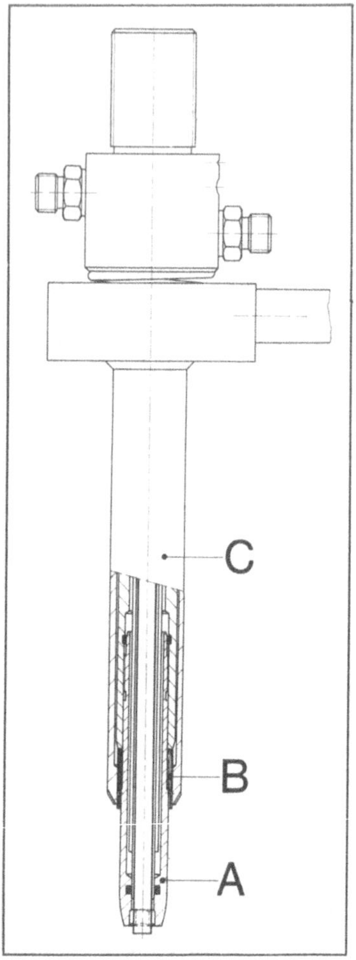

FIG. 10. Typical modern blowpin incorporating a detachable tip. A, Replaceable rotating sleeve for neck flash separation; B, replaceable mandrel tip and cutting sleeve; C, twist sleeve for neck flash separation.

operating process is identical to CO_2 but the temperature of the liquid nitrogen is much lower (−190°C compared with −40°C in the case of CO_2).

3.8.5.3. Hunkar ISC system. Compressed air at 150 atm. is produced with a three-step compressor, passed through an oil separator and filter, and enriched with water for use at a pressure of 70 bar. The 70 bar water/air mixture is injected into the blow moulded container. Due to the reduction of pressure to blow moulding level (approximately 7 bar) the air expands and its temperature is strongly reduced. In addition, the water particles evaporate and pick up heat on the surface of the plastics material and create intensive cooling.

There are some indications that the use of such cryogenic systems produces components with inferior physical properties. More recently the use of small air coolers, fitted to each blow moulder, have shown useful increases in output rates. Blowing air temperature is reduced to around 7°C. As with water cooling of moulds the volume flow of air is most important and the blow pin orifice diameter, in conjunction with the valves, pipework and exhaust mufflers, should be as large as possible.

3.8.6. *Wide Necked or Undercut Containers*

Wide neck containers or bottles with inward turning lips can be produced by using (i) patented 'in-mould cutting' systems (e.g. Bekum) for round, square (see Fig. 11) and rectangular cross-section necks or (ii) out

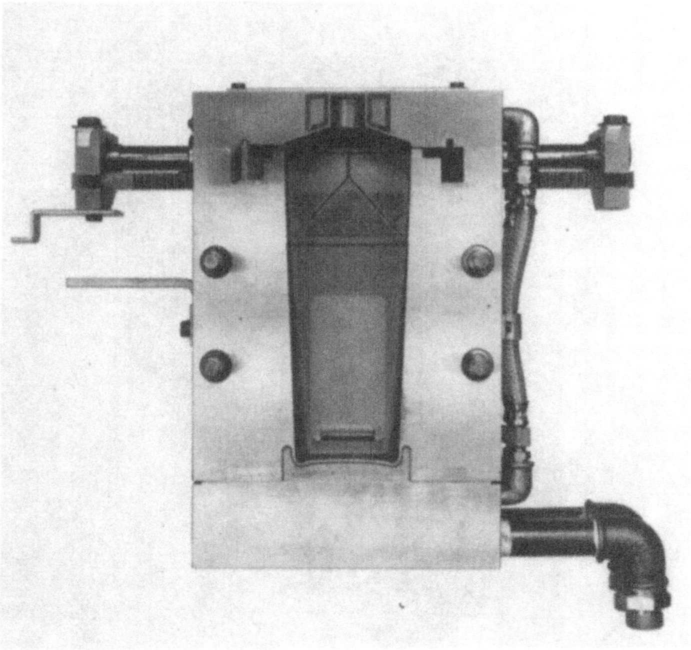

FIG. 11. Blow mould for 1-litre milk carton with automatic square cutting.

of the mould neck trimming systems (either within or outside the machine cycle). Neck designs must suit the trimming system selected. Systems are available from most machine manufacturers. A third alternative for wide necks, especially those with thick cross-sections, is the use of vacuum calibration 'cheeks' sited at the top of the mould, immediately above the neck, to hold the parison in position while the blow pin calibrates.

Containers with undercuts and/or recessed handles can now be blown and deflashed fully automatically by the use of hydraulic mould core movement, and/or by using specially designed undercut punch machines and tools.

3.9. Automation

With increased competition in the trade blow moulding market place, much interest has been generated in automation to reduce labour costs and to improve both production efficiency and product quality. We believe that this trend was established by household brand name com-

FIG. 12. An example of a production line.

panies, determined to keep manufacturing costs to a minimum, and able to do so with their long runs.

An example of a production line is shown in Fig. 12.

Blow moulded products are fully finished either within the machine cycle or alongside the machine and can be automatically leak tested before being conveyed by any of several methods to a point where they can be hand-packed in cartons, tumble-packed in polythene bags, or stored in an intermediate silo. As most in-plant operations print, fill and pack on one shift only, silo storage is finding increasing acceptance as the blow moulders must be operated continuously to achieve acceptable production efficiencies.

The flash or scrap polymer from each component is automatically separated and conveyed to a suitably sized granulator 'beside the press'. The resulting regrind or rechip material is then automatically blended with virgin material and masterbatch (where applicable) via a sophisticated materials handling system and the combined diet fed back to the extruder's hopper.

For most of the smaller machines, hydraulic or pneumatically operated punch deflashing equipment is utilised within the machine cycle. For larger containers, particularly those designed to be 'stackable', punches are now available to remove undercut handle flash automatically.

4. ACCUMULATOR BLOW MOULDING

This process is primarily used for large containers (between 20 and 3000 litres capacity) and for large technical mouldings. An accumulator is used when the cycle time and parison size/weight, combined with the physical strength of the molten polymer, make continuous extrusion impossible. Typically this will be for conventional containers larger than 15 litres in capacity.

4.1. The Process

The extruder feeds molten polymer into an accumulator head and when the required capacity has been reached the parison is extruded very rapidly between the open mould halves (typically 20 kg are ejected in 3–4 s). The parison is normally extended across the bottom by the use of stretchers which are mounted on the calibration station under the mould. Most machines operate on the bottom blow principle.

Extruder design improvements are the same as those mentioned in

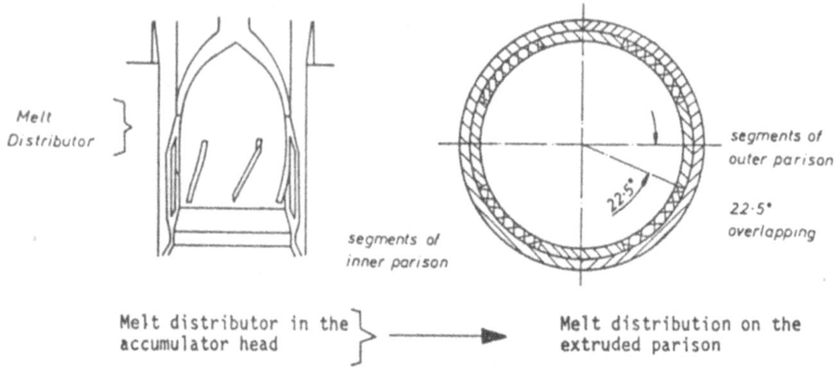

FIG. 13(a). Principle of radial melt distribution.

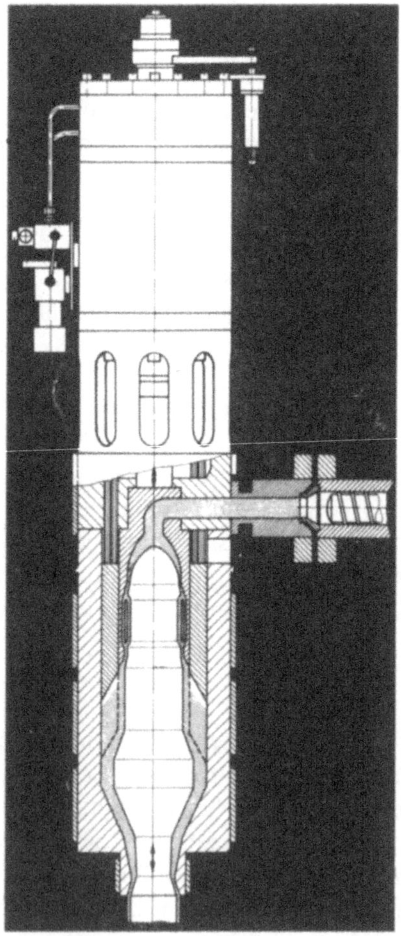

FIG. 13(b). Typical modern extrusion accumulator head design.

Section 3.1 for polyolefins, namely grooved barrels and shear screws. For large containers the trend is towards the use of high molecular weight polyethylene (HMWPE) rather than HDPE.

Accumulator head designs are based on the principle of FIFO (first material in first out), thus ensuring constant heat history. Some heads use melt distributors or vanes (see Fig. 13(a)) to ensure uniform circumferential wall thickness distribution and excellent parison quality at high ejection velocities. Accumulator heads are available with capacities from 2 to 40 litres. Figure 13(b) shows a typical modern accumulator head design.

It is now possible to achieve radial wall thickness and weight tolerances of ±7% and ±1%, respectively, in normal production.

Electronic parison wall thickness controllers are regularly used and have been specially developed to a point where they control the whole machine sequence. The unit generates a start signal for parison ejection when the accumulator is fully charged. Mould closing is then initiated at the end of parison ejection. The filling volume of the accumulator head can be pre-selected digitally. The parison programme points are automatically spread over the volume set.

4.2. General Safety Features

General safety features now used on large machines include visual and audible alarm signals, and fault location by individual indicating lights, for the following malfunctions:

(i) Extruder start-up safety circuit. This provides temperature control of the extruder feeding zone cooling water (for grooved barrel/shear screw extruders only).
(ii) Heater band failure.
(iii) Hydraulic filter blockage.
(iv) Excessive hydraulic oil temperature.
(v) Excessive pressure in accumulator die head.
(vi) Fault in product take-off.
(vii) General cycle fault when pre-set time is exceeded.

4.3. Products

As with extrusion blow machines, manufacturers have developed systems where the containers are automatically transferred by conveyors to punch deflashing units or other finishing stations. In-line container testing is also available. Automatic flash regrind, blending and recycling are also widely used.

New and innovative containers and technical mouldings continue to appear. Recent entries include automotive parts, petrol tanks, traffic light housings, washing machine tubs, L-ring drums, crop spraying tanks, etc.

Special machines have very recently been developed for the petrol tank market by several accumulator machine manufacturers.

5. INJECTION BLOW MOULDING

The process of injection blow moulding has been available for container production for many years and is much more popular in the USA than in Europe. Until recently, only a few UK companies were involved, possibly due to some patent restrictions and the limited availability of suitable machines. Today there are several machine manufacturers with a wide range of machines.

5.1. Materials and Products

Extensive development work carried out by the machine manufacturers, in conjunction with the raw material manufacturers, has led to machines which will process not only the easy materials (polyolefins and polystyrenes), but the more difficult materials such as the polyamides, polycarbonate, acrylonitriles and rigid PVC. PVC injection blow moulding has long been awaited by processors, as it is this material which has found widest application for containers made by extrusion blow moulding, in the cosmetic and pharmaceutical markets. Consequently, there is today much interest in Europe in this process.

Injection blow moulding is not a replacement for extrusion blow moulding, but can best be used for small container production (less than $500\,cm^3$), where the container specification demands exceed the technical, quality and economic limits possible conventionally, i.e. by extrusion. It can be justified economically only if, from the start, it is certain that large quantities, 0·75–1 million containers per cavity, per annum, are required. This is due both to the relatively high tooling costs (typically twice those for extrusion blow moulding) and comparatively high machine capital costs.

5.2. The Basic Process

The basic process consists of three stages.

(a) Injection moulding of a parison/pre-form on to a steel core pin in a hot runner mould.

(b) Transfer of this core pin, with its pre-form, to a blow mould and blowing of the pre-form to the cavity shape.
(c) Transfer of the blown container on its core pin to the ejection station and subsequent ejection.

All machine systems are based on this principle, but differ with regard to the technique used for parison transfer and the method of handling the moulds (rotary table, in either horizontal or vertical operation).

With heat-sensitive rigid PVC there has been steady development. We have gone from one- to two- and recently to four-cavity production. At the moment, development is under way on eight- to 12-cavity production of PVC containers. This should be available within the next few years.

PVC materials have been developed specifically for injection blow moulding and now have little or no impact modifier and a different lubrication system. These changes in formulation mean that lower-cost PVC can be used. The changes are due to the improved mechanical properties inherent in the injection blow process, resulting from the higher degree of orientation achieved.

5.3. Advantages and Disadvantages

The injection blow process has both advantages and disadvantages in relation to the more widely used extrusion blow process. A list of these also provides information on design parameters and limitations.

5.3.1. Advantages

(a) Containers have a high gloss surface finish and are of noticeably improved transparency.
(b) High dimensional accuracy can be attained, particularly in the injection moulded neck area.
(c) There are no welded seams or pinch off scars, and therefore no weak points.
(d) Wall thickness distribution is improved.
(e) Improved weight control, both cavity-to-cavity and shot-to-shot, is possible.
(f) Mechanical properties are improved due to the orientation effect inherent in the process.
(g) Production is flash free. Ancillary equipment, such as granulators and blenders, is not required, saving both on capital and floor space requirements. Additionally, only virgin polymer is used (in extrusion blow moulding of small containers, e.g. 100 cm^3 eye

dropper bottles, flash can be up to 150% of the bottle weight). The elimination of flash also increases production efficiency due to elimination of rejects caused by contamination of flash/regrind.

(h) Production efficiency of operation is increased from 95 to 98%.

(i) Labour requirements are reduced. Machines work fully automatically without need for patrolling by skilled setter/operators.

(j) Lower-cost PVC can be used.

5.3.2. Disadvantages or Restrictions

(a) Mould costs are high, typically twice those for extrusion blow moulding as an injection mould, a blow mould and three sets of core pins are needed for each cavity.

(b) There is only one opening, that produced by removing the core pin.

(c) The process produces centre neck containers only.

(d) No handled or asymmetric containers can be produced.

(e) Container neck diameter-to-length ratio must not exceed 1:10.

(f) The minimum container neck bore is 6–8 mm. The core pin must be stiff enough to resist deflection during injection of the pre-form and large enough to allow adequate passage of the blow air. The bore must also be cylindrical for at least half the inner neck diameter.

(g) High-volume runs are essential for economic reasons.

(h) The maximum container size is 1 litre; normally containers less than 300 cm^3 in volume are produced.

(i) The maximum blow ratio is 4:1, the base punt depth should be between 1 and 1·5 mm, and the width-to-depth ratio on oval or flat containers should be less than 2·5:1.

Over the next few years we expect to see injection blow moulding more widely used for the production of PS, PE, PP and PVC containers in the cosmetic, pharmaceutical and food and drink markets. It will not, with certain exceptions, replace extrusion blow moulded products; rather, it will find new markets for plastic containers.

6. CO-EXTRUSION

This process is a natural development of extrusion blow moulding and the same principles of machine construction, movement and operation apply.

6.1. Reasons for Development

The intention was to introduce containers with improved performance properties, i.e. oxygen permeation, sterilisation, chemical-solvent resistance, scratch resistance and printability, which allow blown plastic containers to enter markets not previously open.

The European market is still limited to special applications due to high polymer costs (particularly the complex adhesive chemicals used to bond the dissimilar polymers), but interest in the new process is high in the toiletry and pharmaceutical areas.

6.2. Container Production

For a three-layer product, the barrier polymer can be the inner or outer layer and results have been successful with nylon and ethylene–vinyl alcohol. The bulk or body polymer is normally one of the polyolefins (LDPE, HDPE or PP). In order to achieve good adhesion between the two layers, a sandwich layer of special polymeric adhesive is also extruded. The machine therefore consists of a main extruder for the polyolefin and two satellite smaller extruders for the barrier material and the adhesive. The extrusion head takes the melt feed from all three extruders and combines them into a single parison. This is then transferred to the mould in the normal way and blown. The three-layer process is now accepted as preferable to the original two-layer (body and adhesive mixed plus barrier) system both economically and technically: economically as much smaller quantities of the expensive adhesive are required, and technically as the potentially dangerous adhesive is sandwiched inside the layers away from contact with pourable liquids.

Reground flash can be recycled carefully as it can sometimes affect the container performance. Regrind can be mixed into the body layer for coloured bottles with a barrier layer on the outside when these are not going to be used for food or pharmaceutical products. If the barrier layer is inside, regrind can be used in the body layer unless it affects the optical properties when natural polymers are in use.

It should therefore be clear that it is necessary to cost containers on 100% virgin materials at a weight including flash. Handled containers and most components that are currently blow moulded can be produced by this process.

7. STRETCH BLOW MOULDING—BIAXIAL ORIENTATION

Stretch blow moulding processes involve the production of a pre-form by extrusion, and extrusion blow moulding or injection moulding while the polymer is in a thermoplastic state. The pre-form can then be cooled and subsequently reheated or tempered before the orientation step. It must be at a precise temperature (in its thermoelastic state), which depends on the polymer being processed. The tempered pre-form is then initially stretched longitudinally followed by radial stretching by blowing the stretched pre-form to the cavity shape. The processes currently available can be used for centre necked, cylindrical conical or oval containers, but not for handled ware. As the materials involved determine the process applied, this section is subdivided according to material.

7.1. Poly (ethylene terephthalate) (PET)

The process applied to this polymer has opened up the very large market for highly carbonated drinks and has in the UK been an area of major capital investment since 1978. Equipment is available from Cincinatti Milacron, Gildermeister Corpoplast/Husky, Sidel and Nissei.

The processes used, with the exception of the one-step process used by Nissei, are two-step involving production of a pre-form (by an injection moulding machine) followed by a second reheating step prior to the stretch blow moulding proper.

7.1.1. Two-Step PET Process

The homopolymer is extremely hygroscopic and very sophisticated drying systems must be used prior to the production of the thick-walled, flanged injection moulding pre-forms. Pre-forms are normally cooled to room temperature and maintained in store for a minimum of 24 h. This curing period is associated with a belief that chemicals which can produce unpleasant odours or flavours in the filled container will diffuse out and away into the air prior to the second step.

The reheat blow moulding machines are large complex devices normally fed by the output from multiple injection moulding machines themselves running with multi-cavity tools. To obtain maximum physical properties in the finished container the thermoelastic temperature range must be very accurately achieved and maintained and this involves a lengthy reheating step. During the whole process the pre-forms are held securely by the flanges below the screw necks on the jars which are a noticeable feature of all PET bottles.

As PET bottles of this type are invariably used for very high CO_2-content drinks, typically 6–8 g CO_2 per litre, the shape of the container must incorporate a heavily domed base section. One or two manufacturers are now producing bottles with blown-in 'feet' although the great majority of bottles are supplied with an injection moulded 'boot' heat-sealed to the base of the container which enables them to be stood upright.

The only problem with this process is the high capital investment involved and the need for continuous runs of simple containers.

7.1.2. One-Step PET Process

Nissei uses the principle of pre-form injection, cooling to intermediate (thermoelastic) stretch blow temperature, stretching and blowing to final form, then finishing cooling all within the same machine frame. With the reservations previously mentioned on chemical tainting, the process offers the obvious advantages of reduced energy consumption, reduced capital cost for a lower-output production unit and facility of supervising the entire moulding operation from granules to packed product.

7.2. Poly (vinyl Chloride) (PVC)

7.2.1. Extrusion Stretch Blow Moulding

This is a continuous process pioneered by Bekum, and is widely used for rigid PVC bottle production from 330 cm^3 to 2 litres capacity. It does not have the ability to contain high CO_2 pressures which makes PET so popular, but its very much lower cost means certain end-uses can still be very efficiently served.

Containers produced by this process exhibit improvements in impact strength, burst strength, resistance to gas permeation and optical properties. All the steps are carried out in sequence on one machine and the bottle is fully finished within the machine frame. The neck can be calibrated or blown with a false top and subsequently 'out of mould' trimmed in-line with the machine.

The above process allows for an increase in container output due to thinner neck sections and the increased cooling time available.

PVC is thermoelastic over a wide temperature range and is, therefore, an ideal polymer for stretch blow moulding. Its formulations do not need high levels of impact modifier for this process: 1% impact modifier or less is required. Thus the raw material is low-cost. (Normal extrusion blow processes use up to 13% modifier to achieve desired impact performance.)

The pre-form is blown in the thermoplastic range, the pre-form is base welded and the flash removed. The neck is either calibrated or blown with a false top. If a calibrated neck is involved, the neck flash is removed. The pre-form is then tempered in the beryllium/copper mould (which has several precisely controlled temperature zones so as to allow an even or profiled temperature) to achieve the required thickness distribution in the finished blown bottle. This zonal temperature control eliminates the need for parison programming in all but the most complex bottle shapes. The tempered pre-form is transferred into the finishing or blowing mould and is initially axially stretched (to 1·3–1·5 times its starting length) by means of a stretch pin; this is followed by additional pre-blowing and it is finally blown and expanded radially into the cavity (2·2–2·7:1).

The base of the finished bottle is also recessed (via a deep punt) to achieve full orientation in the base/side wall area, which is essential for good impact performance. The process is used for both cylindrical and oval containers. For the latter, shaped pre-forms are required. Output rates on twin-head, twin-station machines are up to 2800/h.

7.2.2. *Injection Stretch Blow Moulding*

This process was specially developed by Voith Fischer for PVC bottles of about 0·33 litre capacity and is used for still or lightly carbonated drinks. It is a single-step, single-cavity process, and the machine is based on a conventional injection blow moulder but with additional stations between the injection mould pre-form and the blow mould so as to permit tempering of the pre-form. Output rates are nominally up to 600/h per cavity.

7.2.3. *Market Application*

The markets open to stretch blown PVC bottles are edible oil; water; fruit squash; disinfectants; household chemicals; and carbonated drinks containing up to 3–4 g CO_2 per litre for 1 litre containers, or up to 6–8 g CO_2 per litre for 1/3 litre capacity or less.

7.3. Polypropylene (PP)

Polypropylene homopolymer, although competitively priced when compared with HDPE, has one very severe shortcoming, that of brittle impact performance at low temperature. This property is vastly improved by biaxial orientation: the blow moulding process also improves rigidity, allowing for weight savings and improvement in optical clarity of finished bottles.

It is not possible to use a one-step process and those processes available produce pre-forms from extruded pipe either in one step in-line

(Hercules–Bekum) or two steps (Phillips–Orbit). Both machines/processes are subject to the same shortcomings (involving high capital cost and inflexibility) as the PET equipment described earlier.

7.3.1. Hercules–Bekum Process

This process commences with a pipe extrusion line for very thick-walled pipe. The pipe is vacuum calibrated and cooled and then transported via a loop stand with buffer action to an oven for reheating.

Due to the crystalline structure of polypropylene, the thermoelastic temperature range for orientation, especially for the homopolymer, is very narrow and the tube must be uniformly heated to just beneath the crystalline melting point. To achieve the precise temperature required by PP for good orientation, long reheat times, via multiple passes, are involved in the very large oven. The heated extruded tube is fed continuously into the double-station stretch blow moulding machine where it is pulled, cut, gripped and then stretched axially and radially as it is blown in the mould cavity. The pre-form stretching allows for orientation in the neck and base areas prior to blowing. The bottle is fully finished before ejection from the mould. All flash is fed to a granulator and then blended back to the extruder.

Containers up to 2 litres in capacity as speeds of up to 2000/h can be produced.

7.3.2. Phillips–Orbit Process

In the first of two steps the thick-walled pipe is extruded, fully cooled and cut into pre-form pieces. Then these pre-forms are fixed into pins on a transport chain and reheated in an oven. The heated pre-form is subsequently stretched and blown. The neck is compression moulded by the neck clamps and is, therefore, not orientated. This limits the tube wall thickness and the degree of biaxial stretch attainable. It also controls the ratio of bottle neck to bottle width diameter.

REFERENCE

1. HUNKAR, D., 'Programmable controls. How to get your money's worth', *Plast. Rubber Wkly*, 13 September 1980, 24.

Chapter 6

RHEOLOGY AND DIE DESIGN

R. S. LENK

London School of Polymer Technology, The Polytechnic of North London, UK

SUMMARY

This chapter considers shear flow through dies of various geometries but excludes any treatment of extensional viscosity or elastic phenomena (normal forces, die swell, melt fracture). A comprehensive analysis of flow in both parallel-sided and tapering channels is given, including an indication of how the operating equations are derived. These equations are also summarised. The importance of profile extrusion is compared with that of solid rod extrudates. A section of this chapter is devoted to recent developments for the design of swan-necked dies and others deal with co-extrusion and rotating dies.

1. INTRODUCTION

The parts of the extruder which precede the die essentially serve the following purposes:

(a) supplying polymer or compound to the screw channel;
(b) transport of the material along the screw channel with the screw acting as a pump;
(c) heating the material by conduction from the heating surfaces (the barrel and sometimes the screw itself);

(d) compacting, melting and shearing the material and thus generating frictional heat;
(e) homogenisation with respect to temperature and pressure;
(f) screening to remove minor mechanical impurities and to further ensure that the melt is as homogeneous as possible;
(g) supplying a steady pressure flow to the die.

The die constitutes the final section of the extrusion machine proper. When the melt reaches its entrance there is only a relatively short flow path left before it emerges. However, the length of the die is out of all proportion to the importance of its function—indeed it may be argued that it represents the entire rationale of the extrusion process, namely the conversion of a viscous polymer melt to a linear profile of precisely specified dimensions which in turn may or may not be subsequently reshaped and which is made to freeze into that desired shape by means of a suitable cooling arrangement as soon as it leaves the nozzle.

The volume output Q of an extruder die must clearly be capable of being expressed in terms of three sets of variables:

(i) Machine variables, notably the die geometry.
(ii) Operational variables, notably the applied head pressure which is controlled by the rotation of the screw pump together with the resistance of the die.
(iii) Material variables, notably the melt viscosity under the prevailing extrusion conditions such as the melt temperature and the shear to which the melt is subjected.

The relationship between the volume output and these variables is embodied by a modified Hagen–Poiseuille equation. This equation was discovered empirically and independently more than 100 years ago by the two scientists after which it is named. Working with simple liquids Poiseuille observed that during flow through a cylindrical channel the output is inversely proportional to the channel length and the viscosity of the liquid, but directly proportional to the pressure difference between the channel entrance and channel exit and proportional to the *fourth power* of the radius. Provided that the proportionality constant and the viscosity have been determined it should then be possible to answer questions such as these:

What screw speed should be selected in order to generate the required pressure so as to obtain a specified volume output using a given polymer melt with a given die?

What should be the length of the die of a specified profile so that a desired volume output may be achieved using a given material which is to be extruded under a certain pressure?

What melt viscosity (and hence what temperature) is required so that a given die should yield the required volume output at the selected operating pressure (or vice versa)?

It is the purpose of this chapter to show how such questions may be answered. The key to the answers lies in the derivation and application of suitable equations which are easily manipulated with or without the aid of a scientific calculator. Such an equation will take the form of a Hagen–Poiseuille equation and it will be sufficiently flexible to accommodate other types of liquids than those which its originators used in the beginning, as well as channels of different geometries, both parallel-sided and tapering. Its application will save material, labour and overheads which would otherwise have had to be spent on time-consuming and unnecessary experimentation on a machine which could have been more suitably employed in producing saleable merchandise. To do otherwise is equivalent to insisting on proceeding by pushcart in the jet age.

2. FUNDAMENTALS

It is proposed to confine the treatment of flow through extruder dies to a consideration of the shear viscosity; the effects of extensional flows, secondary flows and normal forces will be ignored even though these are undoubtedly important in certain specialised extrusion operations such as, for example, melt spinning, blow moulding and wire coating. This approach is justified inasmuch as shear viscous phenomena are usually dominant in profile extrusion, which latter accounts for the bulk of extrusion processing; furthermore, there is no reason why the results thus obtained should not be further refined after serving—as they undoubtedly do—as working benchmarks for engineering design calculations.

2.1. Newtonian Fluids

The Hagen–Poiseuille equation in its original form

$$Q = \frac{\alpha \Delta P R^4}{\eta L}$$

is not suitable for polymer melts since it only applies to Newtonian liquids, that is to say, to liquids in which the shear stress τ generated as a result of applying a given shear rate $\dot{\gamma}$ (or vice versa) is linearly related to the latter, the proportionality factor being the coefficient of viscosity, η:

$$\tau = \eta\dot{\gamma}$$

This simple equation is known as Newton's Law and is of the same form—although there are dimensional differences—as Hooke's Law for perfectly elastic solids: $\sigma = E\varepsilon$, where σ is the applied tensile stress, ε the resultant strain (elongation) and E is Young's modulus, a material-specific measure of stiffness or rigidity, i.e. a measure of resistance to deformation in tension. Indeed, the coefficient of viscosity is likewise a modulus of a kind since it is a measure of resistance to flow, but it differs from Young's modulus in that it includes the dimension of time.

2.2. Power Law Fluids

However, polymer melts do not obey Newton's Law except at very low shear, just as many materials do not obey Hooke's Law in the solid state except at very low tensile stresses. A number of rheological equations for liquid flow have been proposed relating τ and $\dot{\gamma}$. The simplest of these is the Power Law

$$\tau = \eta\dot{\gamma}^n$$

where n (the power index) is a number representing the slope of a double-log plot of τ versus $\dot{\gamma}$. The Power Law is a useful tool for the flow analysis of extrusion processes. The power index of polymer melts generally ranges from 0·25 to 0·75, but 0·4 is a commonly occurring value within the temperature range at which the operations are carried out. The value of n is slightly temperature-dependent. It is useful to note that Newtonian behaviour emerges as a special case of the Power Law when n is set equal to unity.

The Power Law serves as a qualitative watershed for distinguishing between shear-thinning ('pseudoplastic') and shear-thickening ('dilatant') liquids; pseudoplastics have power indices of less than unity and most polymer melts fall into this category, whilst dilatants have a power index which is greater than unity. Other flow phenotypes such as 'Bingham plastics' and 'plastics' (not in the polymer sense!) need not normally concern us for the present purpose, but it is worth noting that *all* flow phenotypes can be accommodated in a 'Generalised Flow Theory'.[1]

2.3. Power Law Flow through Cylindrical Channels

A modified Hagen–Poiseuille equation can be derived using the Power Law as the rheological equation in place of Newton's Law. Flow through a cylindrical channel is considered first. Under steady-state conditions the balance of forces is set up as follows:

τ	$\times$ $2\pi RL$	$=$ ΔP	$\times$ πR^2
Shear stress per unit area	Area over which the shear stress is exerted, i.e. the internal die surface	Pressure difference between die entrance and die exit	Area over which the pressure is exerted

It is seen that

$$\tau = \frac{\Delta PR}{2L} = \eta\dot{\gamma}^n = \eta\left(\frac{\partial v_z}{\partial r}\right)^n$$

where v_z is the flow velocity at any radial point r in the forward (z) direction. Making the velocity gradient the subject of the equation and integrating between the limits of $r=R$ (at the die wall, where v_z is zero) and the central axis (where $r=0$ and v_z is at a maximum) one obtains

$$v_z = -\frac{n}{n+1}\left(\frac{\Delta P}{2\eta L}\right)^{1/n} R^{(n+1)/n}\left[1-\left(\frac{r}{R}\right)^{(n+1)/n}\right]$$

the velocity profile of the die for any point along the radial coordinate.

Imagine that the entire cross-sectional area is subdivided into a large number of concentric annuli, each with a mean radius r and a breadth Δr. If the area of each of these annular elements is multiplied by the particular flow velocity v_z that rules over it and if all these products are summed together, then in the limit (when Δr becomes dr) we obtain the volume output Q:

$$Q = 2\pi\int_0^R v_z r\,\mathrm{d}r$$

Substituting for v_z and carrying out the integration between the stated limits we obtain

$$Q = \frac{\pi n}{3n+1}\left(\frac{\Delta P}{2\eta L}\right)^{1/n} R^{(3n+1)/n}$$

We recover the Newtonian case by setting n equal to unity, when it is seen that

$$Q = \frac{\pi \Delta P R^4}{8 \eta L}$$

the original Hagen–Poiseuille equation in which the proportionality constant (in this geometry) is $\pi/8$.

2.4. The Rabinowitsch Correction

Using the last equation together with Newton's Law and substituting for τ as obtained from the balance of forces the Newtonian shear rate $\dot{\gamma}_N$ is found to be equal to $\frac{4Q}{\pi R^3}$.

Proceeding similarly but using the general (Power Law based) Hagen–Poiseuille equation one obtains the true shear rate:

$$\dot{\gamma} = \frac{3n+1}{n} \frac{Q}{\pi R^3}$$

so that it becomes possible to express the true shear rate in terms of the Newtonian (or apparent) shear rate:

$$\dot{\gamma} = \frac{\dot{\gamma}_N}{4} \frac{3n+1}{n}$$

This is the 'Rabinowitsch Correction' which makes it possible to calculate $\dot{\gamma}$ once n and Q have been experimentally determined. This, however, applies to cylindrical channels only; the analogous treatment of rectangular prismatic channels gives

$$\dot{\gamma} = \frac{2n+1}{n} \frac{\dot{\gamma}_N}{3}$$

and

$$\dot{\gamma}_N = \frac{6Q}{WH^2}$$

so that

$$\dot{\gamma} = \frac{4n+2}{n} \frac{Q}{WH^2}$$

For rectangular channels, therefore,

$$Q = \frac{\Delta P W H^3}{4 \eta L} \frac{n}{2n+1}$$

and in the Newtonian case ($n = 1$)

$$Q = \frac{\Delta P W H^3}{12 \eta L}$$

2.5. Tapered Dies

It is convenient to make ΔP the subject of Rabinowitsch–corrected Hagen–Poiseuille equations, no matter what specific die geometry is involved. When extruding polymer melts it has been found to be highly beneficial to use tapering dies since this reduces flow defects; this simple modification makes it possible to operate at shear stresses which are substantially higher than those which mark the limit for obtaining acceptable extrudate in parallel-sided dies. The taper angles are usually less than 10° since the resulting dies would otherwise be too short, but preferably not less than 4° or 5° since the cotangent of the taper angle would then become subject to considerable error. Attention to this has been drawn by Cogswell and Lamb[2] who gave a treatment of the problem and also by Plajer.[3] Lenk and Frenkel[4] have proposed a more rigorous treatment which is, at the same time, remarkably simple and capable of extension to rectangular, elliptical and regular polygonal geometries.

3. ANALYSIS OF STEADY-STATE SHEAR-VISCOUS FLOW IN CHANNELS OF VARIOUS GEOMETRIES

The flow of polymer melts through circular, square, rectangular and triangular dies has been investigated by Ramsteiner.[5] In the following, however, a complete analysis is presented which includes other shapes as well and which also deals with tapering dies.

The method of analysis consists, essentially, in obtaining an expression for the pressure drop ΔP in terms of the channel geometry, the shear-viscous characteristics of the fluid and the volume output. The Power Law is used and care is taken to apply the Rabinowitsch Correction which is appropriate to the particular flow geometry. In order to extend the method from parallel-sided to converging channels an incremental pressure drop $\mathrm{d}P$ along a corresponding channel length $\mathrm{d}l$ is integrated between the limits of a suitable parameter characterising the entrance and exit boundaries. The convergence is linear. In the cases of polygonal and elliptical channels it is assumed that the isovels resemble those of circular and rectangular channels respectively. This assumption is increasingly justified as the number of sides N in a regular polygonal channel exceeds 4 and as the aspect of an ellipse approaches that of a wide slit. In the case of square dies there is a choice of two methods, the rectangular or the polygonal approach, of which the second is perhaps somewhat unnatural and may quite possibly produce less accurate results. In the case of the equilateral triangle the circular approximation is probably subject to considerable error and therefore an alternative 'best guess' type of Rabinowitsch Correction which is thought to be appropriate to that geometry is used. In the event it is seen that the 'error', even with N as low as 3, becomes negligible as the power index n falls below 0·4, a value which is quite common for polymer melts under normal processing conditions.

3.1. General Treatment of Rectangular Channels

3.1.1. Parallel-Sided Channels

The term 'wide slit' is clearly an arbitrary concept, but it is quantified by a shape factor ξ which represents the W/H ratio, the value of which ranges from unity (square) to infinity (wide slit). From the balance of forces it is seen that the shear stress τ is given by

$$\tau=\frac{\Delta P\xi H}{2L(\xi+1)}$$

so that the extreme cases are represented by

$$\tau=\frac{\Delta PH}{2L}\text{ (wide slit)}$$

and

$$\tau=\frac{\Delta PH}{4L}\text{ (square).}$$

Using the Rabinowitsch Correction in rectangular geometries

$$\dot{\gamma}=\frac{2n+1}{3n}\dot{\gamma}_{\mathrm{N}}=\frac{2n+1}{3n}\frac{6Q}{\xi H^3}=\frac{Q}{\xi H^3}\frac{4n+2}{n}$$

and the Power Law $\tau=\eta\dot{\gamma}^n$, it is seen that

$$\eta=\frac{\tau}{\dot{\gamma}^n}=\frac{\Delta P\xi H}{2L(\xi+1)}\left[\frac{\xi H^3 n}{Q(4n+2)}\right]^n$$

and on making ΔP the subject and rearranging,

$$\Delta P = 2\eta L(\xi+1)\left(Q\frac{4n+2}{n}\right)^{n}\xi^{-(n+1)}H^{-(3n+1)} \tag{1}$$

3.1.2. Tapering Rectangular Channels

Consider a tapering channel of constant cross-sectional rectangular shape, i.e. one in which ξ is constant along the entire length L.

The height h at any point along the length l is defined by

$$h = H_1 - 2l\tan\theta$$

and the width $W(=\xi h)$ similarly by $\xi h = \xi H_1 - 2l\tan\phi$, where θ and ϕ are the vertical and lateral taper angles respectively. Eliminating l it is seen that

$$\xi = \frac{\cot\theta}{\cot\phi}$$

and that

$$\mathrm{d}l = -\frac{\mathrm{d}h}{2}\cot\theta \tag{2}$$

From eqn. 1 the incremental pressure drop dP is obtained by substituting dl for L and h for H. Integration between the limits of H_1 and H_2 after substitution for dl according to eqn. 2 yields

$$\Delta P = \frac{\eta\cot\theta}{3n}(\xi+1)\,\xi^{-(n+1)}\left(Q\frac{4n+2}{n}\right)^{n}(H_2^{-3n} - H_1^{-3n}) \tag{3}$$

The matter is somewhat more complicated if ξ is not constant along the entire length. Indeed, it is possible to deal with such a situation if one assumes that the diminution in width $(W_1 - W_2)$ is small compared with W_1. Fortunately most dies conform to this and the assumption allows the binomial theorem to be used. Without giving the detailed working (which is published elsewhere),[4,5] the equation then is of the following form:

$$\Delta P = \left[\eta\cot\theta\left(Q\frac{4n+2}{W_1 n}\right)^{n}\right]\left[\frac{H_2^{-2n} - H_1^{-2n}}{2n}\left(1 + \frac{nH_1\cot\theta}{W_1\cot\phi}\right)\right.$$
$$\left. + (H_2^{1-2n} - H_1^{1-2n})\frac{n}{W_1(1-2n)}\frac{\cot\theta}{\cot\phi}\right] \tag{4}$$

It is seen that L does not appear in either eqn. 3 or eqn. 4, nor does W_2 appear in eqn. 4, since they are already defined by the remaining geometrical parameters.

3.2. Circular Channels

The balance of forces gives

$$\tau=\frac{\Delta PR}{2L}$$

and the Rabinowitsch Correction is

$$\dot{\gamma}=\frac{3n+1}{4n}\dot{\gamma}_{\mathrm{N}}=\frac{3n+1}{n}\frac{Q}{\pi R^3}$$

Using the Power Law, as before, one obtains

$$\Delta P=2\eta L\left(Q\frac{3n+1}{\pi n}\right)^n R^{-(3n+1)} \tag{5}$$

for a cylindrical channel and

$$\Delta P=\frac{2\eta\cot\theta}{3n}\left(Q\frac{3n+1}{\pi n}\right)^n (R_2^{-3n}-R_1^{-3n}) \tag{6}$$

for a truncated right cone of entrance and exit radius R_1 and R_2 respectively and with taper angle θ.

3.3. Regular Polygonal Channels[6]

Let N = the number of sides of length S of a regular polygon; α = the half-angle subtended by one of these sides with respect to any concentric circle ($\alpha=\frac{\pi}{N}$); R = the radius of a circle the area of which is equal to the area of the polygon. Elementary geometry shows that the area of the polygon is $(S^2/4)N\cot(\pi/N)$, so that

$$R=\frac{S}{2}\sqrt{\frac{N}{\pi}\cot\frac{\pi}{N}} \tag{7}$$

From the balance of forces

$$\tau=\frac{\Delta PS}{4L}\cot\frac{\pi}{N} \tag{8}$$

The approximated Rabinowitsch Correction for regular polygons can be shown to be[6]

$$\dot{\gamma}=\frac{3n+1}{4n}\left(\sqrt{\frac{N}{\pi}\tan\frac{\pi}{N}}\right)^{(n-1)/n}\dot{\gamma}_{\mathrm{N}} \tag{9}$$

Equation 9 can be readily checked: for small angles ($N\rightarrow\infty$), tan $\pi/N\rightarrow\pi/N$ and the square-root term becomes unity. This, of course, is the case of the circle and we recover the familiar expression

$$\dot{\gamma}=\frac{3n+1}{4n}\dot{\gamma}_{\mathrm{N}}$$

Working through the arithmetic it is seen that

$$\Delta P=2\eta L\left(Q\frac{3n+1}{\pi n}\right)^{n}\left(\frac{N}{\pi}\tan\frac{\pi}{N}\right)^{n/2}R^{-(3n+1)} \tag{10}$$

where R is a function of S as defined by eqn. 7.

Turning now from a prism to a truncated pyramid and following the same technique as that used for converging channels before, one obtains, analogously:

$$\Delta P=\frac{2\eta\cot\theta}{3n}\left(Q\frac{3n+1}{\pi n}\right)^{n}\left(\frac{N}{\pi}\tan\frac{\pi}{N}\right)^{n/2}(R_2^{-3n}-R_1^{-3n}) \tag{11}$$

Inspection of the ratio λ^n of ΔP values for the polygon and the circle shows that this is equal to

$$\lambda^n=\left(\frac{N}{\pi}\tan\frac{\pi}{N}\right)^{n/2}$$

This ratio λ^n represents the factor of pressure increase required to maintain a given volume flow rate Q through a tapering channel of polygonal cross-section relative to that required for flow through a truncated cone with identical entrance and exit cross-sectional area and with identical length.

The values for N/π (in radians), π/N (in degrees) and $\tan\pi/N$ are listed in Table 1 for polygons with N ranging from 3 to 15. The table also gives the appropriate values for λ^n with n ranging from 0·25 to unity.

TABLE 1
λ^n AS A FUNCTION OF N AND n

N	$\frac{N}{\pi}$ (radians)	$\frac{\pi}{N}$ (degrees)	$\tan\frac{\pi}{N}$	λ^n					
				$n=1$	$n=0{\cdot}75$	$n=0{\cdot}55$	$n=0{\cdot}40$	$n=0{\cdot}33$	$n=0{\cdot}25$
3	0·955	60	1·732	1·288	1·208	1·109	1·042	1·014	1·003
4	1·273	45	1·000	1·128	1·095	1·051	1·020	1·007	1·002
5	1·591	36	0·727	1·076	1·056	1·031	1·012	1·004	1·001
6	1·910	30	0·577	1·050	1·037	1·020	1·008	1·003	1·001
7	2·228	25·71	0·482	1·036	1·027	1·015	1·006	1·002	1·000
8	2·546	22·5	0·414	1·027	1·020	1·011	1·004	1·001	1·000
9	2·864	20	0·364	1·021	1·016	1·009	1·003	1·001	1·000
10	3·183	18	0·325	1·017	1·013	1·007	1·003	1·001	1·000
11	3·501	16·36	0·294	1·014	1·010	1·006	1·002	1·001	1·000
12	3·820	15	0·268	1·012	1·009	1·005	1·002	1·001	1·000
13	4·137	13·85	0·246	1·009	1·007	1·004	1·001	1·000	1·000
14	4·456	12·86	0·228	1·008	1·006	1·003	1·001	1·000	1·000
15	4·774	12	0·213	1·008	1·006	1·003	1·001	1·000	1·000

A plot of these results is presented in Fig. 1.

For Newtonians ($n=1$) it is seen that the 'error' involved in the approximation is as much as 29% for the triangle, 13% for the square and 8% for the pentagon, but thereafter falls rapidly as N increases beyond 5. For pseudoplastics, however, this error reduces substantially as n reaches typical values for polymer melts and at $n=0{\cdot}33$, for example, the error even in the triangle is a mere 1·4%. This suggests that the approximation should be of considerable practical value.

3.4. Alternative Treatment of Triangular Channels

An alternative treatment of triangular channels follows which is not dependent upon the assumptions made earlier, but which is based upon the specific geometry involved.

The balance of forces shows that

$$\tau=\frac{\Delta PS\sqrt{3}}{12L} \tag{12}$$

and, according to Cogswell,[7] the 'best guess' for $\dot{\gamma}_{\mathrm{N}}$ is

$$\dot{\gamma}_{\mathrm{N}}=\frac{Q}{2S^3}\sqrt{\frac{3}{2}} \tag{13}$$

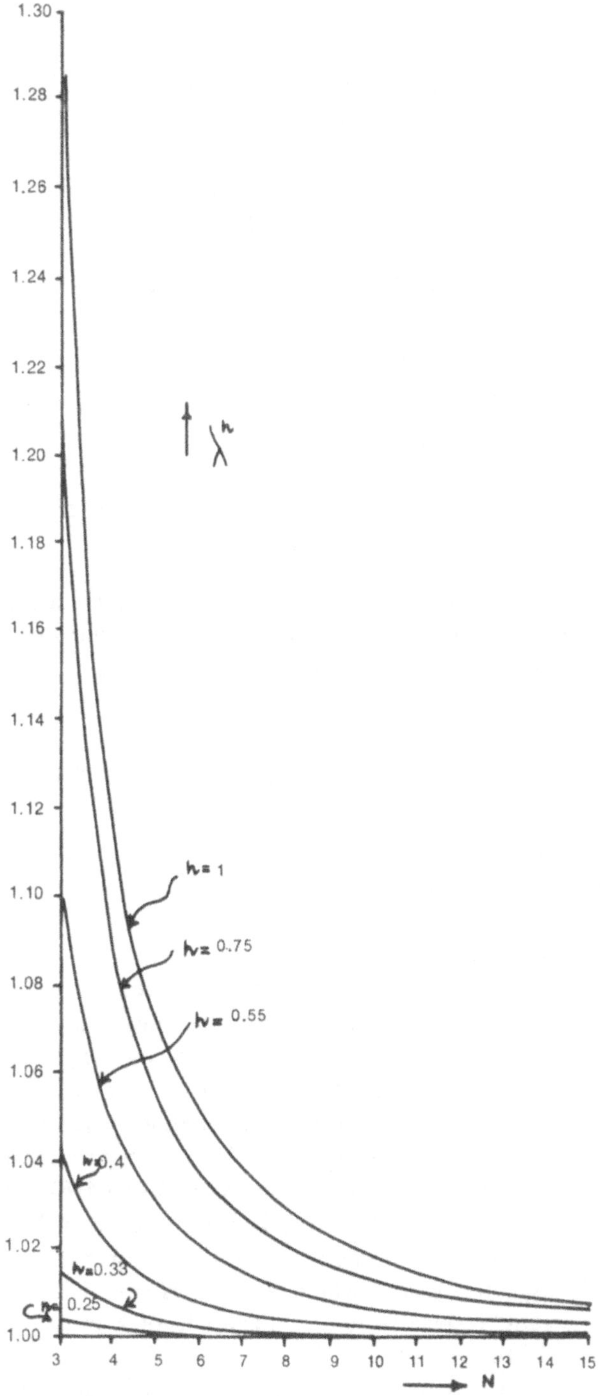

FIG. 1. Plot of λ^n as a function of N and n.

which can then be reasonably approximated by the equation which applies to circular ducts, namely:

$$\dot{\gamma}=\frac{3n+1}{4n}\dot{\gamma}_{\mathrm{N}}=\frac{3n+1}{8n}\sqrt{\frac{3}{2}\frac{Q}{S^3}} \tag{14}$$

Using the Power Law and proceeding as before,

$$\Delta P=4\eta L\left(Q\frac{3n+1}{8n\sqrt{2}}\right)^n 3^{(1-n)/2}S^{-(3n+1)} \tag{15}$$

for a trigonal prism and

$$\Delta P=\frac{3^n\eta\cot\theta}{n}\left(Q\frac{3n+1}{64n\sqrt{2}}\right)^n(H_1^{-3n}-H_2^{-3n}) \tag{16}$$

for the truncated trigonal pyramid, where H_1 and H_2 are the entrance and exit cross-sectional heights of the equilateral triangles respectively. This may, of course, be re-written in terms of S_1 and S_2 rather than H_1 and H_2, giving

$$\Delta P=\frac{\eta\cot\theta}{n}\left(Q\frac{3n+1}{6n}\sqrt{6}\right)^n(S_1^{-3n}-S_2^{-3n}) \tag{17}$$

It should be noted that the integration leading to eqn. 16 was with respect to dh, dl having been replaced by d$h/2\cot\theta$.

3.5. Elliptical Channels[8]

Having previously assumed that the isovels in regular polygonal channels resemble those of circular channels, provided that the number of sides (N) is sufficient, it can be inversely argued that the patterns of isovels in elliptical channels are similar to those of rectangles. Whilst this may not, at first sight, work out too well for relating a square to a circle, it improves with increasing eccentricity of the ellipse which clearly progressively resembles a wide slit as the axial ratio (A/B) increases. The general treatment of rectangular dies has produced eqn. 1.

Now consider an elliptical die in which the major and minor axes are $2A$ and $2B$, respectively, and where $A/B=W/H=\xi$, W and H being the width and height, respectively, of a rectangle of a cross-sectional area equal to that of the ellipse.

Equating the two areas it is seen that $H=B/\sqrt{\pi}$ and the balance of forces gives

$$\tau = \frac{\Delta P \xi B}{L(\xi + 1)} \tag{18}$$

Using the Rabinowitsch Correction for rectangular channels

$$\dot{\gamma} \simeq \frac{Q}{\xi H^3} \frac{4n+2}{n}$$

in combination with the Power Law, it is seen that, for a parallel-sided elliptical channel

$$\Delta P = \eta L (\xi + 1) \xi^{-(n+1)} \left(\frac{Q}{\pi^{3/2}} \frac{4n+2}{n} \right)^n B^{-(3n+1)} \tag{19}$$

Considering a converging elliptical channel of constant shape ($A/B = a/b = \xi$) and taper angle, θ, the usual method (i.e. integration between the limits of B_1 and B_2, and elimination of dl in terms of db and θ) gives

$$\Delta P = \frac{\eta \cot \theta}{3n} (\xi + 1) \xi^{-(n+1)} \left(\frac{Q}{\pi^{3/2}} \frac{4n+2}{n} \right)^n (B_2^{-3n} - B_1^{-3n}) \tag{20}$$

The rectangular approximation is appropriate for 'typical' ellipses. These are defined as ellipses in which $1{\cdot}2 \leq \xi \leq 10$. For ellipses of ξ between 1·0 and 1·2 these may be taken to be, for practical purposes, circles ($\xi = 1$) with radius $(A + B)/2$. At the other extreme ($\xi \geq 10$) the ellipse is so squashed that it is, in practice, just a wide slit and may therefore be treated as such.

3.6. Summary of Flow Equations in Various Die Geometries

(i) *Cylindrical dies*

$$Q = \frac{\pi n}{3n+1} \left(\frac{\Delta P}{2\eta_N L} \right)^{1/n} R^{(3n+1)/n}$$

or

$$\Delta P = 2\eta L R^{-(3n+1)} \left(Q \frac{3n+1}{\pi n} \right)^n \tag{5}$$

Rabinowitsch Correction:

$$\dot{\gamma} = \frac{3n+1}{n} \frac{\dot{\gamma}_N}{4} = \frac{3n+1}{n} \frac{Q}{\pi R^3}$$

(ii) *Convergent circular dies*

$$\Delta P = \frac{2\eta \cot\theta}{3n}\left(Q\frac{3n+1}{\pi n}\right)^n (R_2^{-3n} - R_1^{-3n}) \tag{6}$$

(iii) *Rectangular prismatic dies* ($W \gg H$)

$$\Delta P = \frac{2L\eta}{W^n H^{2n-1}}\left(Q\frac{4n+2}{n}\right)^n$$

Rabinowitsch Correction:

$$\dot{\gamma} = \frac{2n+1}{3n}\dot{\gamma}_N = \frac{2n+1}{3n}\frac{6Q}{WH^2} = \frac{4n+2}{n}\frac{Q}{WH^2}$$

(iv) *Convergent rectangular dies* ($W \gg H$)

(a) *H constant*:

$$\Delta P = \frac{\eta \cot\phi}{n-1} H^{-(2n+1)}\left(Q\frac{4n+2}{n}\right)^n (W_2^{1-n} - W_1^{1-n})$$

(b) *W constant*:

$$\Delta P = \frac{\eta \cot\theta}{2n} W^{-n}\left(Q\frac{4n+2}{n}\right)^n (H_2^{-2n} - H_1^{-2n})$$

(c) *Simultaneous convergence of W and H* (W/H *constant*):

$$\Delta P = \left[\eta \cot\theta\left(Q\frac{4n+2}{W_1 n}\right)^n\right]\left[\frac{H_2^{-2n} - H_1^{-2n}}{2n}\left(1 + \frac{nH_1 \cot\theta}{W_1 \cot\phi}\right) + (H_2^{1-2n} - H_1^{1-2n})\frac{n}{1-2n}\frac{\cot\theta}{W_1 \cot\phi}\right]$$

(v) *Square prism dies* ($W = H = S$)

$$\Delta P = 4\eta L\left(Q\frac{4n+2}{n}\right)^n S^{-(3n+1)}$$

(vi) *Convergent square dies* ($W = H = S$)

$$\Delta P = \frac{2\eta \cot\theta}{3n}\left(Q\frac{4n+2}{n}\right)^n (S_2^{-3n} - S_1^{-3n})$$

(vii) *Rectangular prisms for any value of* $W/H = \xi$ *from very large to* *unity (square), but with ξ constant*

$$\Delta P = 2\eta L(\xi+1)\xi^{-(n+1)}\left(Q\frac{4n+2}{n}\right)^n H^{-(3n+1)}$$

If $\xi=1$, $\Delta P\rightarrow$case (v), if ξ very large, $\Delta P\rightarrow$case (iii).

(viii) *Rectangular converging dies for any value of $W/H=\xi$ from very large to unity (square), but with ξ constant*

$$\Delta P = \frac{\eta\cot\theta}{3n}(\xi+1)\xi^{-(n+1)}(H_2^{-3n}-H_1^{-3n})\left(Q\frac{4n+2}{n}\right)^n$$

If $\xi=1$, $\Delta P\rightarrow$case (vi), if ξ very large, $\Delta P\rightarrow$case (iv).

(ix) *Regular polygonal prismatic dies*

$$\Delta P = 2\eta L\left(Q\frac{3n+1}{\pi n}\right)^n\left(\frac{N}{\pi}\tan\frac{\pi}{N}\right)^{n/2} R^{-(3n+1)} \tag{10}$$

$$\text{where}\quad R=\frac{S}{2}\left(\frac{N}{\pi}\cot\frac{\pi}{N}\right)^{1/2}=\xi_N S \tag{7}$$

and the Rabinowitsch Correction is:

$$\dot{\gamma}=\dot{\gamma}_N\frac{3n+1}{4n}\left(\frac{N}{\pi}\tan\frac{\pi}{N}\right)^{(n-1)/2n} \tag{9}$$

(x) *Regular polygonal converging dies*

$$\Delta P = \frac{2\eta\cot\phi}{3n}\left(Q\frac{3n+1}{\pi n}\right)^n\left(\frac{N}{\pi}\tan\frac{\pi}{N}\right)^{n/2}(R_2^{-3n}-R_1^{-3n}) \tag{11}$$

where $R=f(S)$ as in (ix). Note that

$$\frac{\Delta P_{\text{polygon}}}{\Delta P_{\text{circle}}}=\left(\frac{N}{\pi}\tan\frac{\pi}{N}\right)^{n/2}=\lambda^n$$

where λ represents the factor for the additional pressure that has to be applied in order to maintain the same flow rate through the polygonal die as that through a circular die of equal cross-section, if $n=1$ (Newtonian).

(xi) *Alternative to (ix) for equilateral triangular prismatic dies*

Assuming (*pace* Cogswell) that $\dot{\gamma}_N$ in this geometry is:

$$\dot{\gamma}_N=\frac{Q}{2S^3}\sqrt{\frac{3}{2}} \tag{13}$$

the Rabinowitsch Correction is

$$\dot{\gamma}=\frac{3n+1}{8n}\sqrt{\frac{3}{2}}\frac{Q}{S^3} \tag{14}$$

and

$$\Delta P=4\eta L\left(Q\frac{3n+1}{8n\sqrt{2}}\right)^n 3^{(1-n)/2}S^{-(3n+1)} \tag{15}$$

(xii) *Alternative to (x) for equilateral triangular converging dies*

$$\Delta P=\frac{3^n\eta\cot\theta}{n}\left(Q\frac{3n+1}{64n\sqrt{2}}\right)^n(H_1^{-3n}-H_2^{-3n}) \tag{16}$$

where H is the height of the equilateral triangle.

Expressing H in terms of S:

$$\Delta P=\frac{\eta\cot\theta}{n}\left(Q\frac{3n+1}{n\sqrt{6}}\right)^n(S_1^{-3n}-S_2^{-3n}) \tag{17}$$

(xiii) *Elliptical channels without convergence*

($A/B=W/H=\xi$, where $2A$ and $2B$ are the principal axes and W and H the side lengths of a rectangle of equal cross-sectional area.)

$$\Delta P=\eta L(\xi+1)\,\xi^{-(n+1)}\left(\frac{Q}{\pi^{3/2}}\frac{4n+2}{n}\right)^n B^{-(3n+1)} \tag{19}$$

(xiv) *Elliptical channels with linear convergence* (*taper angle* θ)

$$\Delta P=\frac{\eta\cot\theta}{3n}(\xi+1)\,\xi^{-(n+1)}\left(\frac{Q}{\pi^{3/2}}\frac{4n+2}{n}\right)^n(B_2^{-3n}-B_1^{-3n}) \tag{20}$$

(xv) *General equation for parallel-sided dies*

$$\Delta P=\eta LQ^n f(n,\xi)X^{-(3n+1)}$$

(xvi) *General equation for linearly tapering dies*

$$\Delta P = \eta \cot\theta\, Q^n f(n,\xi)\,(X_2^{-3n} - X_1^{-3n})$$

In the last two equations $f(n, \xi)$ is a function depending on the melt rheology and the geometry; X is a geometrical characteristic of the die having the dimension of a length (e.g. radius); X_1 and X_2 are the entrance and exit equivalents of X.

3.7. Proof that a Parallel-Sided Die Represents Just a Special Case of a Linearly Tapering Die with Zero Taper Angle[9]

It has been shown how the pressure drop in linearly tapering dies can be derived from the pressure drop in parallel-sided dies of the same geometry. Thus, for a circular geometry, we derived eqn. 6 (truncated cone) from eqn. 5 (cylinder). It should equally be possible to recover eqn. 5 from eqn. 6 by setting the taper angle θ to zero. By geometry, tan $\theta = (R_1 - R_2)/L$ or $R_1 = R_2 + L \tan\theta$, so that

$$R_2^{-3n} - R_1^{-3n} = R_2^{-3n} - (R_2^2 + L \tan\theta)^{-3n}$$

Expansion of the second term on the right hand side of this equation, using the binomial theorem, gives

$$(R_2 + L\tan\theta)^{-3n} = R_2^{-3n} + (-3n)\, L\tan\theta\, R_2^{-(3n+1)} + \frac{(-3n)(-3n-1)}{2!} R_2^{-(3n+2)} (L\tan\theta)^2 + \ldots$$

As θ tends to zero, so $\tan\theta$ also tends to zero and one can neglect all the terms containing $(L\tan\theta)^k$, where k is an integer ≥ 2. The last equation then becomes

$$R_2^{-3n} - R_1^{-3n} = R_2^{-3n} - (R_2^{-3n} - 3nL\tan\theta\, R_2^{-(3n+1)})$$
$$= 3nL\tan\theta\, R_2^{-(3n+1)}$$

and substituting in eqn. (6) yields:

$$\Delta P = \frac{2\eta\cot\theta}{3n}\left(Q\frac{3n+1}{\pi n}\right)^n 3nL\tan\theta\, R_2^{-(3n+1)}$$
$$= 2\eta L\left(Q\frac{3n+1}{\pi n}\right)^n R^{-(3n+1)}$$

which is identical with eqn. 5. The same principles can, of course, be equally applied to linearly tapering dies of any other geometry.

3.8. Experimental Evaluation of Flow Equations

Some early results obtained on a Davenport extrusion rheometer[10] using cylindrical, truncated cone and parallel-sided and linearly tapering regular hexagonal dies are given in Tables 2 to 5; the materials investigated were two extrusion grade low density polyethylenes at three temperatures each. It is seen that the calculated pressure drops are, in general, within 20% of the observed entrance pressures, even without making any end-corrections.

TABLE 2

VALUES OF n AND η FOR TWO LOW DENSITY POLYETHYLENES FLOWING THROUGH A CYLINDRICAL DIE

$R=2{\cdot}00$ mm, $L=60{\cdot}00$ mm, piston cross-sectional area = $12{\cdot}5\,\text{mm}^2$. The shear rates were in the range of $10\text{–}60\,\text{s}^{-1}$, corresponding to piston speeds of 1–6 cm/min. The viscosity values η for each polymer grade at each of the three temperatures selected were almost identical within the quoted range of shear rates and the figures given below represent mean values.

Polymer grade	*Temp. (°C)*	n	$\eta \times 10^{-4}$ *(poise)*
Alkathene 33 (MFI=0·3)	190	0·40	2·19
	200	0·44	1·69
	210	0·46	1·42
Alkathene 48 (MFI=0·7)	190	0·43	1·51
	200	0·44	1·37
	210	0·47	1·04

3.9. Examples of Die Design Calculations

3.9.1. Truncated Cone Die: Pressure Requirement

A truncated cone die has the following dimensions:

Orifice diameter $2R_2 = 0{\cdot}40$ cm
Length $L = 5{\cdot}0$ cm
Taper angle $\theta = 5{\cdot}71°$, so that $\cot\theta = 10$

A polymer melt having a viscosity coefficient $\eta = 10^4$ poise and a power index $n = \frac{2}{3}$ is to be extruded at a volume flow rate of $1{\cdot}0\,\text{cm}^3\,\text{s}^{-1}$. It is required to calculate the pressure just inside the die entrance in order to achieve this objective.

TABLE 3

PRESSURE DROP IN A TRUNCATED CONE DIE

$R_1 = 5{\cdot}34$ mm, $R_2 = 1{\cdot}38$ mm, $L = 40{\cdot}00$ mm, $\theta = 5°39'$. Appropriate values for n and η as obtained from a cylindrical die were used (see Table 2).

Polymer	*Temp.* (°*C*)	*Piston speed* (*cm/min*)	$\Delta P_{obs.}$ ($Nm^{-2} \times 10^{-5}$)	$\Delta P_{theor.}$ ($Nm^{-2} \times 10^{-5}$)	$\Delta P_{obs.}/\Delta P_{theor.}$	
Alkathene 33 (MFI = 0·3)	190	1	15·14	12·71	1·19	Mean = 1·29
		2	20·68	16·46	1·26	
		3	25·23	19·46	1·29	
		4	28·53	21·64	1·32	
		5	31·27	23·65	1·32	
		6	33·92	25·44	1·33	
	200	1	13·57	10·90	1·24	Mean = 1·31
		2	18·61	14·27	1·30	
		3	22·33	17·17	1·30	
		4	25·64	19·25	1·33	
		5	28·62	21·48	1·33	
		6	32·51	23·42	1·39	
	210	1	11·17	9·48	1·18	Mean = 1·29
		2	16·13	12·93	1·25	
		3	20·27	15·59	1·30	
		4	23·16	17·67	1·31	
		5	25·81	19·44	1·33	
		6	28·54	21·14	1·35	
Alkathene 48 (MFI = 0·7)	190	1	11·91	9·47	1·26	Mean = 1·24
		2	15·30	12·47	1·23	
		3	18·20	14·89	1·22	
		4	20·68	16·74	1·24	
		5	22·75	18·42	1·23	
		6	25·47	20·06	1·27	
	200	1	10·09	8·95	1·28	Mean = 1·21
		2	13·57	11·79	1·15	
		3	16·54	14·00	1·18	
		4	19·03	15·65	1·22	
		5	21·09	17·39	1·21	
		6	23·00	18·57	1·24	
	210	1	7·78	7·13	1·09	Mean = 1·17
		2	10·75	9·69	1·11	
		3	13·73	11·85	1·16	
		4	16·13	13·56	1·19	
		5	18·20	15·06	1·21	
		6	20·27	16·25	1·25	

TABLE 4

PRESSURE DROP IN A REGULAR HEXAGONAL PRISMATIC DIE

$S = 2{\cdot}20$ mm, $L = 60{\cdot}00$ mm, cross-sectional area $= 12{\cdot}57\ \text{mm}^2$. Appropriate values for n and η as obtained from a cylindrical die were used (see Table 2).

Polymer	*Temp.* (°*C*)	*Piston speed* (*cm/min*)	$\Delta P_{obs.}$ ($Nm^{-2} \times 10^{-5}$)	$\Delta P_{theor.}$ ($Nm^{-2} \times 10^{-5}$)	$\Delta P_{obs.}/\Delta P_{theor.}$	
Alkathene 33 (MFI = 0·3)	190	1	34·16	37·10	0·92	Mean = 0·95
		2	43·68	48·23	0·91	
		3	54·84	56·76	0·97	
		4	59·97	63·13	0·95	
		5	67·25	68·94	0·98	
		6	71·13	74·47	0·96	
	200	1	29·78	32·31	0·92	Mean = 0·91
		2	38·30	42·27	0·91	
		3	46·49	50·86	0·91	
		4	51·29	57·04	0·90	
		5	57·90	63·67	0·91	
		6	64·11	69·41	0·92	
	210	1	25·89	28·30	0·91	Mean = 0·91
		2	34·16	38·61	0·88	
		3	42·19	46·57	0·91	
		4	46·74	52·78	0·89	
		5	52·94	58·05	0·91	
		6	59·14	63·14	0·94	
Alkathene 48 (MFI = 0·7)	190	1	23·58	27·95	0·84	Mean = 0·89
		2	32·10	36·65	0·88	
		3	39·29	43·94	0·89	
		4	43·84	49·41	0·89	
		5	49·63	54·37	0·91	
		6	53·77	59·20	0·91	
	200	1	22·17	26·52	0·84	Mean = 0·85
		2	29·78	34·93	0·85	
		3	35·16	41·48	0·85	
		4	40·54	46·40	0·87	
		5	45·08	51·54	0·82	
		6	48·56	55·04	0·88	
	210	1	18·19	21·40	0·85	Mean = 0·85
		2	24·65	29·04	0·85	
		3	28·95	35·51	0·82	
		4	34·33	40·65	0·84	
		5	38·05	45·14	0·84	
		6	42·44	48·70	0·87	

TABLE 5

PRESSURE DROP IN A TAPERING REGULAR HEXAGONAL DIE

$S_1 = 5{\cdot}87$ mm, $S_2 = 1{\cdot}52$ mm, $L = 40{\cdot}00$ mm, $\theta = 5°39'$. Appropriate values for n and η as obtained from a cylindrical die were used (see Table 2).

Polymer	*Temp.* (°*C*)	*Piston speed* (*cm/min*)	$\Delta P_{obs.}$ ($Nm^{-2} \times 10^{-5}$)	$\Delta P_{theor.}$ ($Nm^{-2} \times 10^{-5}$)	$\Delta P_{obs.}/\Delta P_{theor.}$	
Alkathene 33 (MFI = 0·3)	190	1	15·47	12·97	1·19	Mean = 1·24
		2	21·76	16·78	1·30	
		3	23·82	19·85	1·20	
		4	28·12	22·06	1·27	
		5	29·78	24·12	1·23	
		6	32·92	26·07	1·26	
	200	1	12·82	11·14	1·15	Mean = 1·23
		2	17·54	14·58	1·20	
		3	21·51	17·54	1·23	
		4	24·98	19·66	1·27	
		5	27·96	21·95	1·27	
		6	30·77	23·93	1·29	
	210	1	10·75	9·69	1·11	Mean = 1·17
		2	13·66	13·23	1·03	
		3	18·45	15·95	1·16	
		4	21·09	18·07	1·17	
		5	24·82	19·88	1·25	
		6	27·46	21·62	1·27	
Alkathene 48 (MFI = 0·7)	190	1	11·17	9·68	1·15	Mean = 1·21
		2	14·64	12·68	1·15	
		3	18·20	15·21	1·20	
		4	21·09	17·10	1·23	
		5	23·82	18·82	1·27	
		6	26·47	20·48	1·29	
	200	1	9·93	9·14	1·09	Mean = 1·14
		2	13·48	12·04	1·12	
		3	15·96	14·30	1·12	
		4	18·20	16·00	1·14	
		5	20·93	17·77	1·18	
		6	22·33	18·97	1·18	
	210	1	7·69	7·31	1·05	Mean = 1·14
		2	11·00	9·92	1·11	
		3	13·40	12·12	1·11	
		4	15·72	13·88	1·13	
		5	18·28	15·41	1·19	
		6	20·68	16·62	1·24	

From simple geometry,

$$\cot \theta = \frac{L}{R_1 - R_2}$$

so that

$$R_1 = \frac{L}{\cot \theta} + R_2 = \frac{5{\cdot}0}{10} + 0{\cdot}20 = 0{\cdot}70\,\text{cm}$$

Using the equation for the pressure drop ΔP in truncated cone dies:

$$\Delta P = \frac{2}{3n}\eta \cot \theta \left(Q\frac{3n+1}{\pi n} \right)^n (R_2^{-3n} - R_1^{-3n})$$

$$\Delta P = \frac{2}{3 \times \frac{2}{3}} \times 10^4 \times 10$$

$$\times \left(1{\cdot}0 \times \frac{3 \times \frac{2}{3} + 1}{3{\cdot}14 \times \frac{2}{3}} \right)^{2/3} (0{\cdot}2^{-3 \times 2/3} - 0{\cdot}7^{-3 \times 2/3})$$

$$= 2{\cdot}92 \times 10^6\,\text{dyne cm}^{-2} = 2{\cdot}92 \times 10^2\,\text{kNm}^{-2} \cong 420\,\text{psi}$$

3.9.2. Truncated Cone Die: Land Length

A polymer melt with a viscosity coefficient $\eta = 3 \times 10^4$ poise and a power index $n = 0{\cdot}5$ is to be extruded through a truncated cone die having an orifice radius $R_2 = 0{\cdot}25$ cm and a taper angle $\theta = 9{\cdot}5°$, so that $\cot \theta = 6{\cdot}0$ in such a way that the pressure generated just inside the die entrance is $2{\cdot}21 \times 10^6$ dyne cm^{-2} and produces a volume output $Q = 1{\cdot}0\,\text{cm}^3\,\text{s}^{-1}$. (Note: $2{\cdot}21 \times 10^6$ dyne cm$^{-2} = 2{\cdot}21 \times 10^2$ kN m$^{-2} \simeq 320$ psi.)

The land length, L, of the die can be calculated by using the same equation as in the preceding example. R_1 is first calculated as follows:

$$\Delta P = \frac{2}{3n}\eta \cot \theta \left(Q\frac{3n+1}{\pi n} \right)(R_2^{-3n} - R_1^{-3n})$$

$$2{\cdot}21 \times 10^6 = \frac{2}{3 \times 0{\cdot}5} \times 3 \times 10^4 \times 6 \times \left(1{\cdot}0 \times \frac{3 \times 0{\cdot}5 + 1}{3{\cdot}14 \times 0{\cdot}5} \right)^{0{\cdot}5}$$

$$\times (0{\cdot}25^{-3 \times 0{\cdot}5} - R_1^{-3 \times 0{\cdot}5})$$

$$R_1 = 1{\cdot}45^{1/1{\cdot}5} = 1{\cdot}28\,\text{cm}$$

From trigonometry: $L = \cot \theta\, (R_1 - R_2)$

$$L = 6(1{\cdot}28 - 0{\cdot}25) = 6{\cdot}18\,\text{cm}$$

3.9.3. Divided Annular Die: Output

A polymer melt having a viscosity coefficient $\eta = 10^4$ poise, a power index $n = 0{\cdot}4$ and a density at the processing temperature of $d = 0{\cdot}8\,\mathrm{g\,cm^{-3}}$ is to be extruded through a parallel-sided die of uniform section thickness. The axial length of the profile die is 6·0 cm and it has the shape of a circular annulus with a straight cross bar running through the centre, thus dividing the tube cavity into two equal and almost hemicircular channels. The outer and inner radii of the annulus are $R_o = 1{\cdot}105$ cm and $R_i = 0{\cdot}805$ cm respectively. What weight output in $\mathrm{kg\,h^{-1}}$ is to be expected if the pressure ΔP generated just inside the die entrance is 10^6 dyne $\mathrm{cm^{-2}}$ (or $10^2\,\mathrm{kNm^{-2}} \cong 140$ psi)?

The weight output (in $\mathrm{kg\,h^{-1}}$) can be calculated as follows.

This die may be treated as a slit, i.e. as a rectangular channel in which the effective slit width W is represented by the sum of the mean circumference of the annulus plus the width of the cross bar:

$$W = 2\pi\bar{R} + 2R_i, \text{ where } \bar{R} = \tfrac{1}{2}(R_o - R_i)$$
$$\therefore W = \pi(R_o - R_i) + 2R_i = 3{\cdot}14(0{\cdot}805 + 1{\cdot}105) + 2 \times 0{\cdot}805 = 7{\cdot}61\,\mathrm{cm}$$

The slit height H is obviously $(R_o - R_i) = 1{\cdot}105 - 0{\cdot}805 = 0{\cdot}3$ cm. The ratio

$$\frac{W}{H} = \xi = \frac{7{\cdot}61}{0{\cdot}3} = 25{\cdot}367$$

Using the equation:

$$\Delta P = 2\eta L(\xi + 1)\left(Q\frac{4n+2}{n}\right)^n \xi^{-(n+1)} H^{-(3n+1)}$$

and putting in the numerical data we solve for Q as follows:

$$10^6 = 2 \times 10^4 \times 6 \times (25{\cdot}367 + 1) \times Q^{0{\cdot}4} \times$$
$$\left(\frac{4 \times 0{\cdot}4 + 2}{0{\cdot}4}\right) \times 25{\cdot}367^{-(0{\cdot}4+1)} \times 0{\cdot}3^{-(3 \times 0{\cdot}4 + 1)}$$

$$Q^{0{\cdot}4} = 0{\cdot}856$$
$$Q = 0{\cdot}678\,\mathrm{cm^3\,s^{-1}}$$

To convert the volume output into the weight output in the required units, Q must be multiplied by the density and by 3600/1000:

$$Q_w = 0{\cdot}678 \times 0{\cdot}8 \times 3{\cdot}6 = 1{\cdot}95\,\mathrm{kg\,h^{-1}}$$

3.9.4. Truncated Cone Die with Mandrel

An annular tube die having the overall shape of a truncated cone is obtained by inserting a conical core of the same length and the same angle of convergence into a hollow truncated cone so that the two surfaces are parallel and a flow channel is generated which is characterised by a constant distance between those surfaces along the entire length of the die. By taking the necessary measurements of a paper model of about the desired size the following die dimensions were established:

	Entrance radius (cm)	*Die-lip radius (cm)*
Inner surface	$(R_i)_1 = 1{\cdot}241$	$(R_i)_2 = 0{\cdot}939$
Outer surface	$(R_o)_1 = 1{\cdot}560$	$(R_o)_2 = 1{\cdot}241$

Overall length $L = 14{\cdot}4$ cm

It is intended to extrude a polymer melt of viscosity coefficient $\eta = 2 \times 10^4$ poise and power index $n = 0{\cdot}35$ through this die at such a rate that the linear take-off velocity of $0{\cdot}5\,\mathrm{cm\,s^{-1}}$ produces an extrudate of exactly the same circumference as that of the outer die lip, thus compensating for die swell. The pressure just inside the die entrance which must be applied in order to achieve this circumference may be calculated in the following way.

The flow channel may be treated as a wide slit. This is justified because, as will be seen presently, W_1 and W_2 are both very much greater than H.

W_1, the mean circumference of the entrance annulus $= 2\pi R_1$

$$= \pi[(R_o)_1 + (R_i)_1] = 8{\cdot}80\ \mathrm{cm}$$

Similarly, W_2 is given by $\pi[(R_o)_2 + (R_i)_2] = 6{\cdot}85$ cm, while the height H of the slit is equal to either $(R_o)_1 - (R_i)_1$ or $(R_o)_2 - (R_i)_2$. These compute to 0·319 and 0·302 respectively, but considering the channel as of constant height, a mean value of 0·31 is taken.

When the flow channel is 'unrolled' and flattened out its plan view is that of a circle sector from which a smaller concentric sector has been removed. The lateral taper angle ϕ can be calculated, but it is conveniently found using the paper model: a chord is drawn between the inlet corners and a perpendicular is drawn from the exit corner to the chord. A right angle triangle is thus generated whose hypotenuse is the outer edge of the slit (i.e. the side length of the truncated cone): ϕ is the thin wedge angle of this triangle. The ratio $\cot\phi$ is found to be $14{\cdot}2/1{\cdot}9=7{\cdot}474$, whence $\phi=7{\cdot}62^\circ$.

The Rabinowitsch-modified Power Law version of the Hagen–Poiseuille equation for the pressure drop in laterally converging wide slits of constant height is

$$\Delta P=\frac{\eta\cot\phi}{n-1}H^{-(2n+1)}\left(Q\frac{4n+2}{n}\right)^n(W_2^{1-n}-W_1^{1-n})$$

Before applying this equation the correct value for the volume output, Q, must be found. Assuming that there is no change of shape of the extrudate due to die swell by appropriately adjusting the draw-down rate, then $Q=A\,v_z$, where A is the cross-sectional area at the die lips and v_z is the given take-off velocity:

$$A=\pi[(R_o)_2^2-(R_i)_2^2]=3{\cdot}14(1{\cdot}54-0{\cdot}882)=2{\cdot}067\,\text{cm}^2$$

$$\therefore\ Q=2{\cdot}067\times0{\cdot}5=1{\cdot}0335$$

We now make the numerical substitutions in the pressure drop equation:

$$\Delta P=\frac{2\times10^4\times7{\cdot}474}{0{\cdot}35-1}\times0{\cdot}31^{-(2\times0{\cdot}35+1)}\times\left(1{\cdot}0335\frac{4\times0{\cdot}35+2}{0{\cdot}35}\right)^{0{\cdot}35}$$

$$\times\,(6{\cdot}85^{1-0{\cdot}35}-8{\cdot}8^{1-0{\cdot}35})$$

$$\Delta P=2{\cdot}33\times10^2\,\text{kN}\,\text{m}^{-2}\simeq330\,\text{psi}$$

The pressure drop as calculated above is of the order of magnitude commonly used in profile extrusion. Had it been an order of magnitude greater, then doubts would have arisen as the whether the tubing can be produced without flow disturbances causing unacceptable defects in the extrudate even if the required pressure can be generated without overtaxing the power capacity of the machine or seriously overstraining the channel through which the melt passes from the screw to the nozzle. The simplest way of coping with such a problem is to change the viscosity coefficient of the melt. Since this is highly temperature-dependent,† it is

†Viscosity/temperature functions and shift factors which make it possible to predict the actual temperature necessary to reduce the viscosity coefficient to a suitable level have been established (see, for example, ref. 11). When such data are available experimental work on the die is reduced to a few runs in order to verify the prediction and perhaps make some slight adjustments in order to compensate for the errors incurred in the approximations used when the flow equation was derived.

possible to reduce it by about an order of magnitude by increasing the temperature by a few tens of degrees, provided that the material will not suffer any structural changes as a consequence; alternatively, one may think about using another polymer grade, say a copolymer or the same polymer containing a little more plasticiser, provided that the shape-holding of the extrudate during cooling can be maintained. Since any change in temperature also slightly affects the power index an iteration of the calculation should be carried out, and this also applies to a change in the polymer grade run at the same temperature.

If, however, neither the die geometry nor the processing temperature can be changed and the polymer grade cannot be altered, then it will be necessary to accept whatever output rate can be obtained on the given machine at the nearest possible pressure to that calculated which still yields an extrudate of acceptable quality. Yet another device would be to introduce a vertical taper (opening the gap from 0·31 to 0·4 cm at the entrance) when the bilateral taper so produced would still retain a virtually constant slit shape ratio W/H and to use the flow equation for that particular geometry.

3.9.5. Blow Moulding Die

A parison forming die for extrusion blow moulding consists of a truncated cone with a core that converges at an angle such that the annular gaps at the die entrance and at the die-lips H_1 and H_2 are 2 cm and 0·5 cm, respectively. By selecting a mean entrance radius of $R_1 = 4{\cdot}775$ cm and a mean exit radius of $R_2 = 1{\cdot}194$ cm, the mean circumferences of the entrance and exit annuli are $W_1 = 30$ cm and $W_2 = 7{\cdot}5$ cm respectively. It is therefore seen that W_1/H_1 $(=30/2) = W_2/H_2$ $(=7{\cdot}5/0{\cdot}5) = 15$ along the entire channel. This means that the channel is a wide slit of diminishing size but of constant cross-sectional shape. When 'unrolling and flattening' a paper model of the die it has a plan view appearance of a 'cloak' (i.e. of a converging pie sector from which a smaller concentric pie sector has been cut away). In order to keep the die fairly short it is necessary to make the subtending angle α of the 'cloak' quite large and it is therefore fixed at 120°. Using an end-corrected reservoir pressure ΔP of $1{\cdot}8 \times 10^6$ dyne cm^{-2} (i.e. $1{\cdot}8 \times 10^2$ kN m^{-2} or $\simeq$ 260 psi), a polymer melt having a viscosity coefficient $\eta = 5 \times 10^4$ poise and a power index $n = 0{\cdot}3$, then the mean linear flow velocity of the extruded tube can be calculated by assuming that die swell and parison sag cancel each other out.

The cotangents of the two taper angles of the wide slit die are first calculated. The lateral taper angle ϕ is clearly half the subtending angle of the pie (α), i.e. 60°, so that $\cot\phi = 0{\cdot}5774$. The vertical taper angle θ is defined by $\cot\theta = A/(H/2)$, where A is the radial distance from the pie centre and H is the corresponding height of the slit at any point A. Since the pie sector angle is 120° the periphery of the outer 'cloak' margin ($= W_1 = 30$ cm) is also equal to one-third of the circumference of the *whole* pie, so that the pie radius is given by: $A_1 = 3W_1/2\pi = 14{\cdot}324$ cm. Similarly, A_2, the radius of the smaller concentric pie sector (which has been cut away) is given by $A_2 = 3W_2/2\pi = 3{\cdot}581$, where $W_2 (= 7{\cdot}5$ cm) is the periphery of the *inner* 'cloak' margin.

Having seen that $\cot\theta = A/(H/2) = \text{constant}$,

$$\cot\theta = \frac{A_1}{H_1/2} = \frac{14{\cdot}324}{2/2} = \frac{A_2}{H_2/2} = \frac{3{\cdot}581}{0{\cdot}50/2} = 14{\cdot}324$$

(and hence θ is 3·99°.)

The Rabinowitsch-modified Power Law form of the Hagen–Poiseuille equation for linearly tapering wide slit channels with simultaneous lateral and vertical tapers and constant cross-sectional shape is now applied.

$$\Delta P = \left[\eta\cot\theta\left(Q\frac{4n+2}{W_1 n}\right)^n\right]\left[\frac{H_2^{-2n} - H_1^{-2n}}{2n}\left(1 + \frac{nH_1\cot\theta}{W_1\cot\phi}\right)\right.$$

$$\left. + (H_2^{1-2n} - H_1^{1-2n})\frac{n}{1-2n}\frac{\cot\theta}{W_1\cot\phi}\right]$$

where $\Delta P = 1{\cdot}8\times10^6$ dyne cm^{-2} $= 1{\cdot}8\times10^2$ kNm^{-2} $\simeq 260$ psi

$\eta = 5\times10^4$ poise	$H_1 = 2$ cm
$\cot\theta = 14{\cdot}324$	$H_2 = 0{\cdot}5$ cm
$n = 0{\cdot}3$	$\cot\phi = 0{\cdot}5774$
$W_1 = 30$ cm	

Introducing these numbers:

$$1{\cdot}8\times10^6 = \left[5\times10^4\times14{\cdot}324\times Q^{0{\cdot}3}\times\left(\frac{4\times0{\cdot}3+2}{30\times0{\cdot}3}\right)^{0{\cdot}3}\right]$$

$$\times\left[\frac{0{\cdot}5^{-2\times0{\cdot}3} - 2{\cdot}0^{-2\times0{\cdot}3}}{2\times0{\cdot}3}\times\left(1+\frac{0{\cdot}3\times2{\cdot}0\times14{\cdot}324}{30\times0{\cdot}5774}\right)\right.$$

$$\left. + (0{\cdot}5^{1-2\times0{\cdot}3} - 2{\cdot}0^{1-2\times0{\cdot}3})\times\frac{0{\cdot}3}{1-2\times0{\cdot}3}\times\frac{14{\cdot}324}{30\times0{\cdot}5774}\right]$$

$Q^{0 \cdot 3} = 1 \cdot 94$
$Q = 9 \cdot 106 \, \text{cm}^3 \, \text{s}^{-1}$
This is the *volume* flow rate. The linear flow velocity is obtained from $v_z = Q/X$, where X is the area of the annulus at the die lips:

$$X = \pi[(R_o)_2^2 - (R_i)_2^2]$$

$$(R_o)_2 = R_2 + \frac{H_2}{2} \qquad \text{and } (R_i)_2 = R_2 - \frac{H_2}{2}$$

$$= 1 \cdot 194 + \frac{0 \cdot 5}{2} \qquad = 1 \cdot 194 - \frac{0 \cdot 5}{2}$$

$$= 1 \cdot 444 \qquad = 0 \cdot 944$$

so that $X = 3 \cdot 14 \, (1 \cdot 444^2 - 0 \cdot 944^2) = 3 \cdot 14 \, (2 \cdot 085 - 0 \cdot 891) = 3 \cdot 751$

$$v_z = \frac{Q}{X} = \frac{9 \cdot 106}{3 \cdot 751} = 2 \cdot 428 \, \text{cm} \, \text{s}^{-1}$$

The result of this calculation is the key to predicting output. Assuming that this die is used to blow mould cylindrical bottles 21 cm high and 7 cm in diameter, a parison length of, say, 25 cm is required allowing for extra length to grip the ends before the inflation. Thus it will take about 10 s to extrude. Since closing the mould, inflation and cooling are quite fast it is clear that it should take about 15 s to make one bottle, so that one could produce at a rate of 240 per hour. It is also possible to calculate the average wall thickness from the draw ratio of parison to the cavity wall in the hoop direction.

4. ENTRANCE CORRECTIONS

4.1. Principles of End Corrections

The flow equations seem to indicate that the pressure drop and the die length are directly proportional. It is certainly true that, in general, the shorter the die the lower is the pressure required to achieve a given volume output. Bagley[12] carried out extrusion experiments with dies of decreasing L/R ratio and plotted L/R versus ΔP. A straight line was obtained with an intercept on the ΔP axis which showed that there was a pressure drop even in a die of zero length; moreover, this pressure drop was by no means negligible. In many cases it is a fair approximation to make a 'Bagley end correction' which assumes that, in a circular die, the pressure drop in a die of zero length is equivalent to an imaginary die length of six radii. This must therefore be added to the *nominal* die length to give the *effective* die length. To obtain a *true* shear stress and a *true* pressure drop in a circular die one must therefore multiply the nominal shear stress and the nominal pressure drop with an empirical factor of $(L/R)/[(L/R)+6]$. This is also useful in allowing for the difference between the L/R ratio of the rheometer used in determining the rheological parameters of the melt on the one hand and the L/R ratio of the proposed die on the other. Obviously, the shorter the die, the more significant will the end correction be. Similar end corrections may be applied to wide slit dies, using a ratio of $2L/H$ in place of L/R.

4.2. Example of the Use of End Corrections

It is proposed to use a die with an L/R ratio of 10·5. The rheometer die, on the other hand, on which the rheological measurements were obtained, had an L/R ratio of 18·3. The *effective* shear stress (and hence the effective pressure drop, using the Bagley end correction) would therefore be obtained by multiplying the *nominal* shear stress by a factor of 0·666 for the proposed die and 0·753 for the rheometer die respectively. These two figures are obtained from the respective ratios of $18{\cdot}3/(18{\cdot}3+6)$ and $10{\cdot}5/(10{\cdot}5+6)$. The appropriate Hagen–Poiseuille equation indicated that a die with 3·175 mm radius and delivering 8·2 cm^3 of extrudate per second (i.e. 24 kg h^{-1}), under an applied pressure of 84·37 kp cm^{-2} would need to be 33·5 mm long. This is the *nominal* die length L_{nom}. Without an entrance correction this would be too long, resulting in a lower volume output, unless the pressure is increased correspondingly. If one wishes to work at the pressure for which the process was originally designed, L_{nom} will have to be multiplied by a factor of $b(a+6)/a(b+6)$, where $a=L/R$ of the rheometer die and $b=L/R$ of the uncorrected die. In the specific case the corrected die length came to 28·5 mm. If the end correction had been neglected the die would have been 18% too long. This would have meant that the volume output would have been a mere 14 kg h^{-1} instead of the intended 24 kg h^{-1}. In order to restore the intended volume output the pressure would have to be raised to 100 kp cm^{-2} from the specified figure of 84·37 kp cm^{-2}. This might well cause undesirable side effects (feed problems, degradation, etc.) to the severe embarrassment of the designer who has failed in his assignment to produce a die which fulfils the requirements of the specified processing conditions.

Further worked examples are given by Carley.[13,14]

5. THE PRACTICAL IMPORTANCE OF WIDE SLIT DIES

Wide slit dies are the most important types of dies that the trade extruder ever handles. This becomes immediately obvious when it is realised that annuli are wide slits which have been bent round an external axis until the ends meet. Furthermore, there is absolutely no need for such profiles to be circular; they may be rectangular or polygonal in shape, L-shaped, V-shaped or U-shaped, or indeed a combination of all these shapes ('lollipops', 'wheels with spokes', etc.), but they are still 'wide slits' as far as their flow geometry is concerned.[15] Thus they include piping, tubing, sliding channels for doors and windows, strip gasketing for weatherproofing and some of the complex shapes required for the snap-fit box sections of extruded PVC window frames, interlocking panelling, extruded skirting board which simultaneously serves as conduit for electrical wiring, and the like, as well as dies for sheet and film (including layflat film) and dies for wire coating. Solid rod dies are used for producing elastomeric compression gasketing, monofilament, fibre and fibrillated string, all of which, though important, between them represent a much smaller volume proportion of all the polymer material that is processed by melt extrusion.

6. FLOW THROUGH CIRCULAR SECTION CHANNELS WITH A 'SWAN NECK' TYPE CONVERGENCE

6.1. Design Criteria for the Die Entrance

While tapers are undoubtedly highly effective in improving flow, it was considered that an ultimate optimisation could only be attained by streamlining the die entrance region as much as possible. This requires the selection of a suitable function which is essentially conicocylindrical and which defines the distance of the wall from the axis at each point in terms of any given distance along that axis. The axis itself is positioned so as to give the desired entrance and exit radii. Lenk and Sláter[16] consider that the die has its optimum shape when the following criteria are met:

(i) The tangents to the die wall at the entrance and exit must be parallel to the central axis.
(ii) The longitudinal course of the die wall (i.e. the radius/length

function) must be sigmoid and antisymmetrical about a point midway between the entrance and exit—that is to say, it has a point of inflection at the origin of the coordinate system in Fig. 2.

(iii) The tangent to the die wall at the midpoint shall be defined by a taper angle θ which is less than 10°. Since the problem requires a numerical solution (as will be seen presently), just three taper angles are arbitrarily selected such that $\cot\theta$ has values of 6, 10 and 20. Solutions are obtained for these angles only, rather than continuously varying values within that range of taper angles.

6.2. Design Solution

It is considered that the length/radius function is a quintic of the general form

$$r = al^5 + bl^4 + cl^3 + dl^2 + el + f \tag{21}$$

in which the coefficients b and d are taken to be zero in order to satisfy condition (ii) above. This function should have a symmetrical shape as depicted in Fig. 2, and this imposes certain limitations on the values of a,

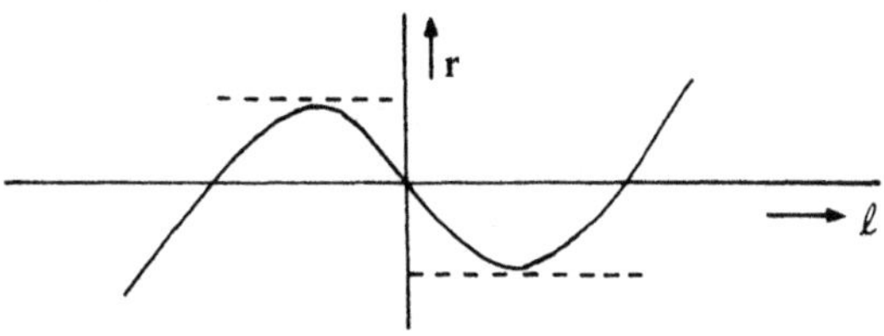

FIG. 2. Centre portion of a symmetrical quintic curve. (Reproduced from reference 16, with permission John Wiley & Sons, Inc.)

c, e and f. It is seen that this function has a maximum and a minimum. At the origin the tangent to the curve has *not* a zero slope. This point is nevertheless considered a point of inflection and it is stipulated that its slope ($\cot\theta$) should have one of the three numerical values stated above, namely 6, 10 or 20.

Taking the centre portion (i.e. the portion between the maximum and the minimum only) and rotating it around an axis parallel to the tangents at the maximum and minimum, a channel is generated with a circular cross section along its entire length, of which the radius r at any point l between entrance and exit is given by eqn. 21. The entrance and exit radii R_1 and R_2, the overall die length L, and the slope m at the conventional point of inflection are obviously related. In practice R_1 and R_2 will be fixed by the screw/barrel and extrudate design respectively, and the overall die length L will be determined by the slope m. A die of this type is diagrammatically shown in Fig. 3.

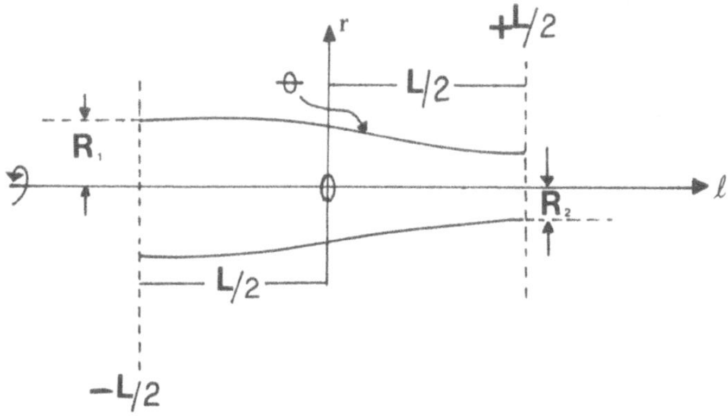

FIG. 3. Schematic layout of die generated from the symmetrical quintic curve. (Reproduced from reference 16, with permission John Wiley & Sons, Inc.)

In the streamlined dies which are now considered the approach is entirely analogous to that used in section 3.2, but difficulties arise in solving the corresponding integral

$$\int_{-L/2}^{+L/2} r^{-(3n+1)} \mathrm{d}l \tag{22}$$

that is,

$$\int_{-L/2}^{+L/2} (al^5 + cl^3 + el + f)^{-(3n+1)} \mathrm{d}l \tag{23}$$

This integral cannot be evaluated by analytical methods, but it can be solved numerically and the method is fully described in Lenk and Slater's paper.[16] Values of 0·1, 0·6 and 1·0 were taken for R_2 and for each of these R_1 was taken such that $R_2 = 0{\cdot}1\ R_1$, $0{\cdot}5\ R_1$, and $0{\cdot}7\ R_1$. Then, in each of these cases, m was taken as 6, 10 and 20. The evaluation of eqn. 23 has been performed for $n = 0{\cdot}3$, 0·5 and 0·7. The value of

$$2\left(\frac{3n+1}{\pi n}\right)^n m \int_{-L/2}^{+L/2} r^{-(3n+1)} \mathrm{d}l = -\frac{\Delta P}{\eta Q^n}$$

was found and graphs were produced of $-\Delta P/\eta Q^n$ versus n for each channel shape.

6.3. Application of the Solution

Curvilinearly tapering dies will certainly be very much more expensive to produce than rectilinear ones, but they would have the considerable advantage of causing a minimum of flow disturbance. The channel is generated by rotating that portion of a quintic function of length and radius around an axis parallel to the tangent to the curve which is between the maximum and the minimum of the curve. Such a die would minimise the flow discontinuity at the entrance where the flow changes from a broad and sluggish to a narrow and fast stream, as well as achieve the benefits accruing from tapered dies generally. All that is necessary is to solve the integral of $f(r)\,\mathrm{d}l$ and this can be done by means of numerical computerisation.

7. CO-EXTRUSION

7.1. Applications

Co-extrusion is a highly sophisticated and fairly recent method for processing plastics, but it has been used in the manufacture of rubber hose for several decades. It has become rapidly established in many sectors of the extrusion field, but the dominant outlet is in packaging, especially for film and sheet. Non-packaging applications include sheet (for the vacuum forming or thermoforming of refrigerator liners, luggage, boats and potentially baths and sinks) as well as pipe, wire coating, cable covering, tubing, hose, profile, fibre and monofilament. Co-extrusion can lead to an entirely novel range of products with special performance characteristics, products which can, moreover, combine this very acceptable feature with considerable cost savings. An example of this would be the combination of virgin material with reground scrap: the latter would be hidden by the virgin material which constitutes the functionally important top surface from the point of view of appearance, wear, chemical resistance, weather resistance, etc.

Co-extrusion has been used for producing 'conjugate' fibres which consist of two semicircular components. The product resembles natural wool due to the crimping which results from differential thermal contraction (and consequent filament buckling) on cooling from the melt.[17-19]

7.2. Equipment

Co-extrusion equipment is generally limited to three systems:[20]

(i) feedblocks, in which melt streams from different extruders are combined in a small cross-section before entering the die;
(ii) multi-manifold internal combining dies, in which the various melt streams enter the die separately and join just inside the die orifice;
(iii) multi-manifold external combining dies with completely separate manifolds for the different melts as well as distinct orifices through which they leave the die in order to join together just beyond it.

7.2.1. General Comments

For blown film, multi-manifold internal combining dies are preferred, mainly because there is some risk of disturbing the laminar flow of the layers when using feedblocks.

Feedblock dies are generally used when heat-sensitive materials are being processed, but such dies are not suitable for handling materials whose viscosities are substantially different (viscosity ratios of 2:1 to 4:1 are said to be the limit), or whose processing temperatures differ significantly (i.e. by about 30°C or more). When the viscosity or temperature differences are large one should use a multi-manifold die; a multi-manifold internal combining die is particularly useful when the temperature differences between the different materials are large. Since the effective shear rate changes continuously in a feedblock it is important that the viscosities of the materials should also change in a comparable manner lest the original material balance be lost.

However, for pipe and tubing the relative cheapness and simplicity of feedblock systems outweigh their disadvantages, although *any* system can, in principle, be used (see Section 7.2.3). When co-extruded film, sheet or coatings are produced by the flat die process there is little to choose between the various systems and any one of those mentioned above may be used. For wire and cable coating, however, multi-manifold dies are used exclusively.

7.2.2. Blow Moulding

Intermittent extrusion blow moulding, using independent concentric accumulators and annular piston rams, is a novel technique for producing multi-layer containers.[21] Figure 4 shows a three-material, three-layer co-extrusion head. The basic design is similar to continuous co-extrusion

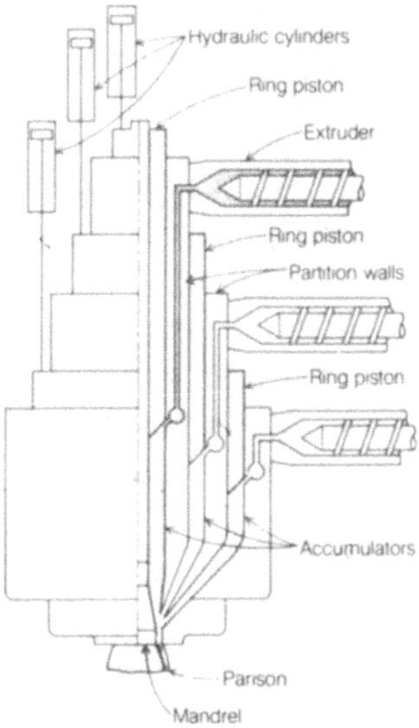

FIG. 4. Three-material, three-layer co-extrusion accumulator head. (Reproduced from reference 21, with permission McGraw Hill, Inc.)

heads—except for the enlarged accumulating sections and corresponding ring pistons. These pistons are the key to intermittent co-extrusion. The plasticated melts are accumulated within separate concentric reservoirs, each with its own independent temperature control system. The parison is formed against the mandrel by forcing the melt from the reservoirs by means of the ring pistons. Each of these is operated by an independent hydraulic cylinder; this allows the extrusion rate (and hence the pressure) of the various melts to be accurately controlled within wide limits, and feeding of one or more layers may be abruptly terminated whenever this is desired. Such on/off control is important for two reasons:

(i) expensive polymers can be positioned within the parison so that they are present in the container body, but not in the flash where they would be wasted;
(ii) the flash, consisting of a single material only, is easily recycled.

7.2.3. Pipe Production

Co-extrusion of pipe has long been recognised as a cost-effective method for the manufacture of useful products, but it has been held back by technical difficulties. For example, in a conventional pipe the central torpedo is supported by a spider and the spider legs introduce weld lines into the finished product. Such weld lines constitute a source of potentially serious mechanical weakness, especially in pressure pipes.

In one recently announced system marketed by Durapipe Ltd[22] the spider is replaced by an adjustable mandrel. The standard design of the die enables one to use three different materials for co-extrusion, but dies for a smaller or larger number of materials can be supplied according to customers' requirements upon request.

Each material is admitted to the head via control valves (see Fig. 5; one is shown at A as fitted to the manifold block of the secondary extruder, B) and it then passes to the circular plenum chambers (C) and thence on to the die.

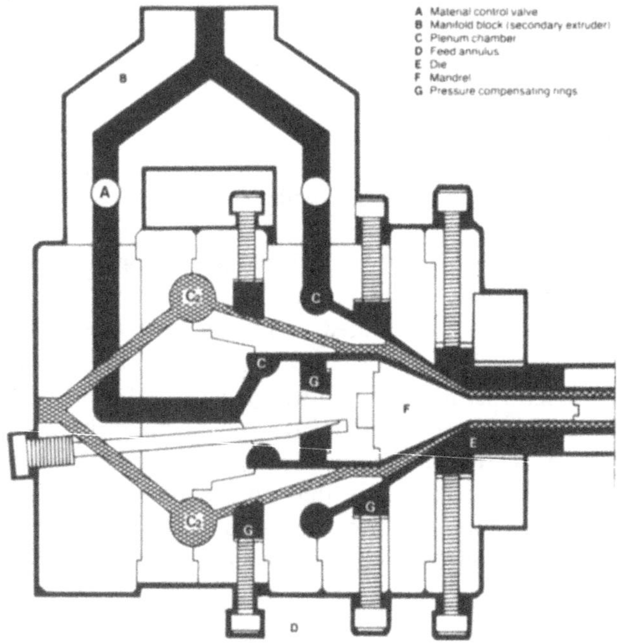

FIG. 5. Durapipe co-extrusion die with plenum chamber feeds and concentricity adjuster bolts. (Reproduced from p. 281 of reference 23, by permission of Academic Press Inc.)

The layer thickness is controlled by adjusting the appropriate valve in the feed manifold. The concentricity of each layer is controlled by adjusting the pressure compensating rings via the seat belts, thus varying the cross-section through the feed annulus to match the pressure drops.

7.3. Rheology of Co-extrusion

When two incompatible liquids of different viscosities are made to flow together then it is often found that there is a reduction in the minimum pressure gradient required to produce flow. This effect has long been exploited in petroleum technology where the addition of a little water to the oil improves the pumping efficiency of the oil in the pipes.

A 'pressure gradient reduction factor' (PGRF) is defined by

$$\left(\frac{\partial P}{\partial z}\right)_{AB} \Bigg/ \left(\frac{\partial P}{\partial z}\right)_{A}$$

where A is the more viscous and B the less viscous component. It was seen (p. 265 of reference 23) that the PGRF remains virtually constant over wide ranges of volume-percentage compositions of the melts—presumably so long as there is sufficient of the lower viscosity component to ensure all-round lubrication. Using a Power Law model and a wide slit geometry, Han (pp. 264–70 of reference 23) derived equations for the volume output of each component as a function of the variables normally involved in single-phase extrusion and additionally as a function of the position of the interface in the cross-sectional die profile.

Since the less viscous phase is the one which flows most actively or easily it will also be the one which preferentially wets the die wall and will eventually envelop the more viscous component. In order that this state should become fully developed it is necessary that the dwell time of the co-extruding liquids in the die be sufficiently long. This is more likely to be achieved when the channel is long and/or when the applied pressure is low or moderate. Han and Khan[24] confirmed this by studying the effect of length/diameter (L/D) ratios of circular dies on the interface shape of polystyrene (PS) and low density polyethylene (LDPE) at a number of different output rates. They found that, after an initial flat interface of two semicircular streams at the die inlet, the more viscous PS phase was completely enveloped by the LDPE at the chosen output rate, provided that the L/D ratio exceeded a certain minimum value. They also carried out a theoretical study for determining the equilibrium interfacial configuration during stratified flow in rectangular ducts and provided theoretical backing for the view that viscosity will be far more important in determining the interfacial geometry than elastic effects.

Han[23] also co-extruded PS and LDPE through a circular die both concentrically and eccentrically. Little change occurred during extrusion when the PS was the core material; but when the LDPE was placed at the centre initially a dramatic—if gradual—reversal could be observed provided that the die was long enough for steady-state conditions to be fully developed.

The viscosity/shear stress plots of individual components of a two-component system are sometimes found to cross, so that the material with the higher viscosity below a certain shear stress becomes the one with the lower viscosity above that shear stress threshold.

Such is the case in the system PS/HDPE (polystyrene/high density polyethylene). At 200°C the viscosity of PS exceeds that of HDPE below approximately $1{\cdot}2 \times 10^6$ dyne cm^{-2} (see Fig. 6(a)); but above that shear

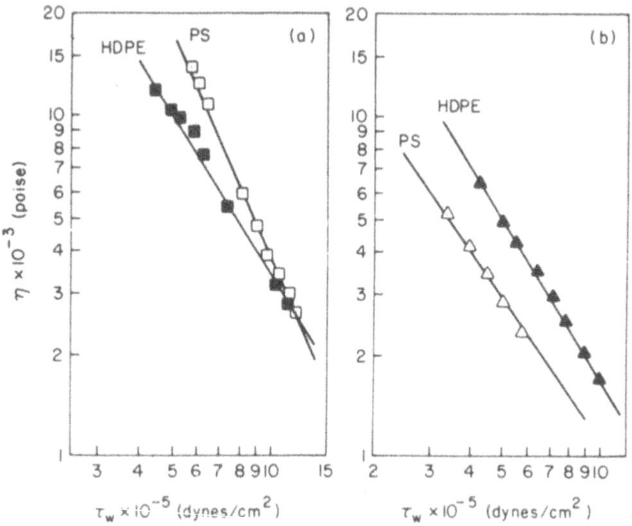

FIG. 6. Viscosity versus shear stress for PS (polystyrene) and HDPE (high density polyethylene), used for co-extrusion experiments. Temperature: (a) 200°C; (b) 240°C. (Reproduced from p. 265 of reference 23, with permission Academic Press, Inc.)

stress HDPE is the more viscous of the two melts. On the other hand, HDPE is the more viscous component over the entire available shear stress range at 240°C (see Fig. 6(b)). When co-extruding at 200°C at a shear stress below $1{\cdot}2 \times 10^6$ dyne cm^{-2} the core material is PS, but when the temperature is raised to 240°C a phase reversal is found to occur resulting in the encapsulation of the HDPE by the PS.

Lee and White[25] and Southern and Ballman[26,27] used polymer systems which gave viscosity crossovers as the shear stress was increased and they duly observed that the envelope reversal also occurred at the same points.

A theoretical prediction of the pressure gradients during the co-extrusion of non-Newtonian fluids has been the subject of a paper by Han and Chin.[28]

7.4. Analysis of Multilayer Flow

A reasonably simple flow analysis is possible in channels with a uni-directional flow such as cylindrical in wide slit dies.

Consider the velocity profile in a multilayer melt system flowing through a slit die of height H as shown in Fig. 7.

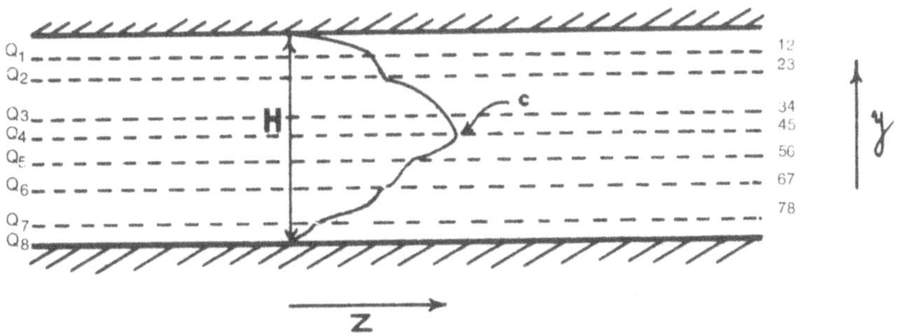

FIG. 7. Flow analysis for multilayer extrusion. (Reproduced from reference 29, with permission SPE.)

Each layer is parallel to that of the major die walls such that the co-extruding polymers alternate and form interfaces 12, 23, etc. The velocity profile consists of essentially parabolic segments and it is assumed that there is no slippage at the interfaces so that the velocity at each interface is the same for each polymer along that plane. The shear stress τ is continuous across the interfaces. Neglecting all components of τ other than τ_{yz}, the shear stress is

$$\tau = \frac{\partial P}{\partial z}(H - c)$$

where c is the distance from the die wall of the plane C where $\tau = 0$. In a symmetrical multilayer $c = H/2$.

Using the Power Law, $\dot{\gamma}_i = \eta_i \tau^{n_i}$, it is noted that the suffix i stands for the parameter appropriate to each layer species and that $\dot{\gamma}_i$ and τ may be either positive or negative. Taking absolutes

$$|\dot{\gamma}_i| = \eta_i |\tau|^{n_i}$$

and specifically

$$\dot{\gamma}_i = \text{sign}(\tau)\eta_i|\tau|^{n_i} \tag{24}$$

where sign $(\tau) = +1$ for $\tau > 0$
sign $(\tau) = -1$ for $\tau < 0$

Since $\dot{\gamma}_1 = \dfrac{\partial (V_i)_z}{\partial y} = \text{sign}\left(\dfrac{\partial P}{\partial z}\right) \text{sign}\,(H-c)\eta_c \left|\left(\dfrac{\partial P}{\partial z}\right)\right|^{n_i} \left|(H-c)\right|^{n_i}$

the velocity profile is obtained by integration between the limits of each layer in turn, the total number of layers being N. For the Mth layer

$$V_M(h) = -\text{sign}\left(\frac{\partial P}{\partial z}\right)\sum_{j=1}^{M-1}\left\{\frac{\eta_j\left|\left(\frac{\partial P}{\partial z}\right)\right|^{n_j}}{n_j+1}[|a_{j-1,j}-c|^{n_j+1} - |a_{j,j+1}-c|^{n_j+1}] + \frac{\eta_M\left|\left(\frac{\partial P}{\partial z}\right)\right|^{n_M}}{n_M+1}[|a_{M-1,M}-c|^{n_M+1} - |H-c|^{n_M+1}]\right\}$$

between the interfaces of that Mth layer.

The volumetric flow rate of that layer is obtained by integrating that equation in turn giving:

$$|Q_M| = \sum_{j=1}^{M-1}\frac{\eta_j\left|\left(\frac{\partial P}{\partial z}\right)\right|^{n_j}}{n_j+1}(a_{M,M+1}-a_{M,M-1})\{|c-a_{j-1,j}|^{n_j+1} - |c-a_{j,j+1}|^{n_j+1}\} + \eta_M\frac{\left|\left(\frac{\partial P}{\partial z}\right)\right|^{n_M}}{n_M+1}(a_{M,M+1}-a_{M,M-1})$$
$$\{|c-a_{M-1,M}|^{n_M+1}\} + \eta_M\frac{\left|\left(\frac{\partial P}{\partial z}\right)\right|^{n_M}}{(n_M+1)(n_M+2)}\{|c-a_{M,M+1}|^{n_M+2}$$
$$\text{sign}(c-a_{M,M+1}) - |c-a_{M-1,M}|^{n_M+2}\,\text{sign}(c-a_{M-1,M})\} \tag{26}$$

If $\left|\dfrac{\partial P}{\partial z}\right|$, c, a_{12}, a_{23}, a_{34} ... $a_{N-1,N}$ are known the equation can be solved.

In practice Q_M is known by experiment and what is wanted is $(\partial P/\partial z)$, the position of the zero-shear plane C and the interfaces a_{12} ..., etc. If an arbitrary value for $(\partial P/\partial z)$ and for c is selected then the equation can be solved for a_{12}, a_{23} ..., etc. in turn and values for $V_M(h)$ obtained. According to Schrenk and Alfrey, Jr.,[29,30] 'the calculated value of $a_{N,N+1}$ does not match the true location and the calculated velocity at this interface does not match the true value (zero)'. These mismatches are removed by iteration with fresh arbitrary values for $(\partial P/\partial z)$ and c and the correct solution is ultimately obtained by convergence.

If the multilayer system is symmetrical the problem is simpler because it is known from the start that $c = H/2$ and one only needs to iterate from an initially assumed value for $(\partial P/\partial z)$ to remove the mismatch with the boundary at H. This, then, affords the complete velocity profile, the locations of the fluid interfaces, and the value for τ at each interface.

The rheology of polymer blends and dispersions has been analysed by Van Oene.[31]

8. ROTATING DIES

8.1. Combined Pressure and Drag Flow

Consider a combination of pressure flow and drag flow where the latter has been superimposed upon the former. Numerous examples of such situations exist. They include extensional flows such as the convergent interacting flows at a die entrance, drag flow during wire coating, leakage flow at the flight tops of a rotating extruder screw, spreading disc flow during the filling of an injection mould cavity, and flow through an annulus with a rotating mandrel. Interpretations and predictions regarding the rheological behaviour in such situations are liable to deviate considerably from the idealised situation which is thought to obtain in a capillary rheometer.

In an extruder screw the reduction in apparent viscosity η_{app} at the top of the flight due to the high rotational shear rate against the barrel facilitates leakage flow and has a lubricating effect.

8.2. Rotating Mandrels

Considering extrusion using a rotating mandrel where a rotational drag flow is superimposed upon the pressure flow, the total shear rate $\dot{\gamma}_{total}$ is given by

$$\dot{\gamma}_{\text{total}} = (\dot{\gamma}^2_{\text{extr}} + \dot{\gamma}^2_{\text{rot}})^{\frac{1}{2}}$$

where

$$\dot{\gamma}_{\text{extr}} = \frac{3Q}{\pi R H^2} \text{ and } \dot{\gamma}_{\text{rot}} = \frac{2\pi R \Omega}{H}$$

(R is the mandrel radius, H the annular gap, Q the volume flow and Ω the angular velocity of rotation).

A wide slit geometry applies, with $\tau = \Delta PH/2L = \dot{\gamma}_{\text{extr}}\eta_{\text{app}}$, η_{app} being the apparent viscosity which corresponds to $\dot{\gamma}_{\text{total}}$. Experimental verification of this approach has been sought. Using a moulding grade of PMMA($T = 230$°C, $\dot{\gamma}_{\text{extr}} = 3{\cdot}4\,\text{s}^{-1}$, $L = 30$ mm, $H = 0{\cdot}5$ mm, $R = 9{\cdot}75$ mm) and superimposed values of $\dot{\gamma}_{\text{rot}}$ of up to $340\,\text{s}^{-1}$, Cogswell[7] found that the calculated values for the pressure drop were in good agreement with those observed, the former showing an error of less than 10% on the low side.

It has long been known that the properties of thermoplastic (and especially glass fibre reinforced) pipes are improved by rotating the mandrel of the extrusion die since this induces a circumferential fibre orientation (and, by implication, orientation of the polymer molecules as well) and thus raises the hoop strength and modulus of the pipe at the expense of the axial strength and modulus. At the same time, it improves the weld strength by modifying the isovels near the weld lines caused by the flow deflector of the die and by the spider legs, especially in fibre reinforced pipes where the fibres do not cross the weld interface.

Die systems with rotating mandrels have been investigated by Worth.[32] It has been suggested[33] that die rotation reduces the power consumption during melt extrusion owing to the additional shear which reduces the viscosity of the non-Newtonian melt. However, this only holds good when the torque applied is very low since, at higher speeds of rotation, the power input via the mandrel far outweighs the saving in extruder power.

Taking the shear rate in the annular die as

$$\frac{\partial v}{\partial y} = \left(\frac{\tau}{\eta}\right)^{1/n}$$

this can be resolved into axial and tangential components

Axial $$\frac{\partial v_z}{\partial y} = \left(\frac{\tau}{\eta}\right)^{1/n} \cos\phi$$

and

Tangential $$\frac{\partial v_x}{\partial y} = \left(\frac{\tau}{\eta}\right)^{1/n} \sin\phi$$

ϕ being the angle of molecular orientation. Integration of these equations and the use of the appropriate boundary conditions enables one to obtain the axial and tangential velocity profiles. In this connection it is of more than passing interest to note that fibre filled polymer melts show characteristic orientation patterns during extrusion, and particularly so when using rotating dies. Although such fibres are generally too short to produce a significant mutual interaction due to entanglement they *can* be thought of as constituting a visualisation pattern, or model, for the way in which molecules will tend to align themselves under the prevailing conditions, although the latter are, of course, very much subject to restrictions due to chain entanglement.

The tangential orientation is at a minimum at the inner and outer surfaces and is complete in the mid-position. Since the relative flow rate in the axial and circumferential directions governs the flow orientation, the optimum overall strength (i.e. the isometric strength) is associated with a characteristic angle of orientation. Experience in continuous filament winding with glass reinforced polyester pipes would suggest that an angle of orientation relative to the axis is optimised around 55°, but it must be borne in mind that the angle in rotating mandrel extrusion—as distinct from continuous filament winding—varies throughout the thickness layers of the pipe.

8.3. Rotating Co-extrusion Dies

Many-layered tubular blown film can be obtained by co-extrusion using rotating dies.[26] Figure 8 shows the effect of die rotation on the pattern of layer generation in an annular die. The actual layer thickness depends on the die length and on the relationship between the velocity of rotation and the extrusion rate. Multi-microlayers, each down to a few hundred Ångstroms thick, have been obtained. In the extreme case the layers may become discontinuous ('layer scission'). When that happens, streaks with broadening ends are formed due to molecular relaxation; further rupture may ultimately produce spherical particles. A similar pattern is observed during the co-extrusion of polymer blends through fibre spinning dies[34] as shown in Fig. 9.[34]

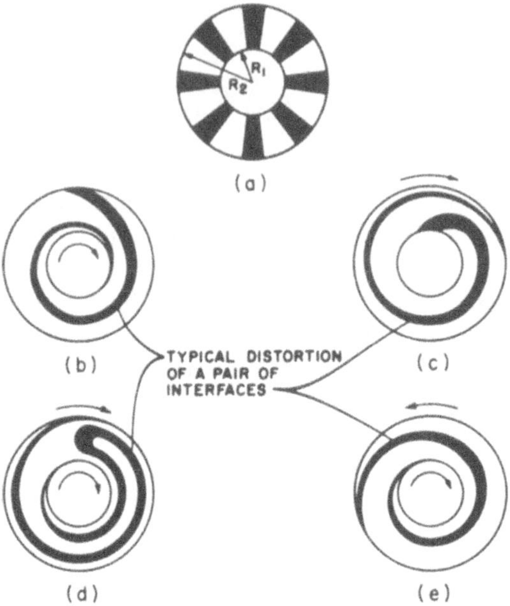

FIG. 8. Four modes of rotation of annular die boundaries generating different layer patterns: (a) feed system distribution entering annular die; (b)–(d) distortion patterns produced. (Reproduced from reference 29, with permission SPE.)

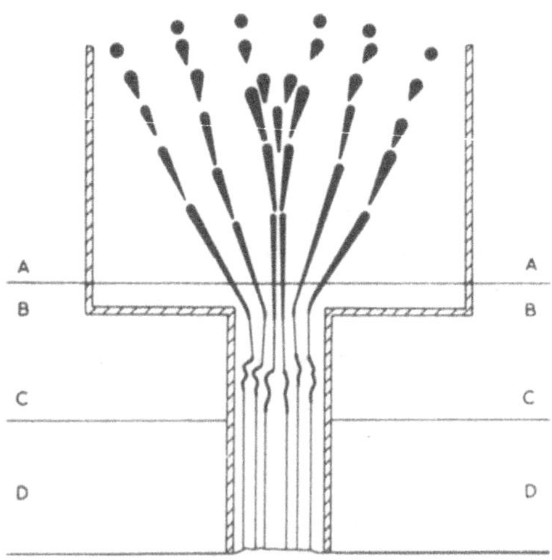

FIG. 9. Empirical model for fibrillar texture formation in a molten polyolefin blend while under tensile stress resulting from convergent flow: A, reservoir; B, primary fibrillation zone with C, 'snapback' (elastic recovery) effects evident; the fixing of the fibrillar texture; D, melt flow with fibre orientation. (Reproduced from reference 14, with permission SPE.)

9. CONCLUSIONS

It is useful first to ask some pertinent questions to which the answers are largely implicit in the aforegone.

9.1. Elongational Flows and Elastic Effects

How can one justify the neglect of elongational flows and the associated pressure drops, as well as the neglect of secondary flows and elastic forces in extrusion dies, despite the fact that they all contribute to the overall pressure drop?

9.1.1. Elongational Flows

Firstly, it is pointed out, once again, that this chapter concerns flow in the die only. In parallel-sided dies at normal extrusion pressures (say, of the order of 10^6 dyne cm^{-2}) it is a reasonable first approximation that the flow velocity profile within the die is constant from end to end, that no extensional flow exists and that the pressure drop within the die is entirely accounted for by the transducer reading just inside the die entrance. (If the transducer reading is made just *outside* the die entrance, then a Bagley correction should be used, especially when the length/diameter ratio (or its equivalent in non-circular geometries) is small.) When there is a *small* taper in the die the resulting rather gradual change in the down-channel velocity profile can still be reasonably explained by the pressure drop equations in their integrated forms. Pressure losses due to elongational flow and due to secondary flows certainly do occur in the transition region upstream of the die and this has to be treated separately, just as the flow analysis in the screw channel and in the intermediate approach region (screen pack, spider legs, etc.) have to be treated in their own right.

9.1.2. Elastic Effects

Normal (elastic) forces may be appreciable in the die inlet region and cause corresponding pressure losses. It is only when we consider the *total* energy balance throughout the entire flow path of the melt through the machine that we need to account for each contribution. But in this context we are only concerned with the die.

Laminar flow in the die may be upset if the extrusion pressure is excessive, but this is generally of the order of 10^6 dyne cm^{-2} (10^2 kN m^{-2} or 10^2–10^3 psi). Not so in injection moulding, where the flow through runners, sprue and during mould filling itself may reach values which are one and even two orders of magnitude higher. Extensional flow then assumes considerable importance, but again, this is not our present concern.

9.2. Secondary Flows

A potential source of error in most design calculations arises from the assumption that the flow is entirely uniaxial. This is well approximated in circular, polygonal and near-circular elliptical dies, but in rectangular dies there is some sideways oscillation (and possibly even some 'porpoising' (up-and-down) flow), except when the rectangle is a wide slit. The viscosity coefficient in bi- and tri-axial flow is really a mean value which may not be the same as the uniaxial viscosity coefficient as determined by extrusion rheometry in which a circular (i.e. a uniaxial) die is usually employed. Carley[14] suggests that the viscosity coefficients as determined using a circular and wide slit die are related by the ratio $\eta_{\circ}/\eta_{\square} \cong 0{\cdot}91$. But since it is common practice to use viscosity data from a standard circular rheometer die as a basis for die design calculations an error of about 10% must be expected when the extrusion die is in fact a wide slit. The engineer can live with this and maybe use an empirical correction factor if he wants to improve on the accuracy.

9.3. Converging Dies

Elongational flows and secondary flows do play some part in converging dies, but this may be neglected when the taper angles are small. We have taken some liberties in this respect with the parison forming blow moulding die (see example in Section 3.9.5) when the *lateral* taper angle ϕ was large lest the overall length of the truncated conical annulus be too long. This is probably permissible as a first approximation.

9.4. Overall Expected Accuracy of Die Calculations

It should not be forgotten that the Power Law itself is only an approximation although it generally works reasonably well.

Looking at all the potential sources of error implicit in the exclusive use of a Rabinowitsch-modified and Power-Law-based Hagen–Poiseuille equation, it is really quite remarkable that the margin of discrepancy between prediction and observation is usually less than 20%, which is acceptable for first-approximation engineering design purposes and which can be improved upon by means of empirical adjustments.

9.5. Reasons for Co-extrusion

What is the point of co-extruding different polymer melts?

The main reason is the desire to produce laminated structures in which discrete layers contribute different properties to the product. This is illustrated by means of three examples:

(i) A film sandwich of nylon 610 between two layers of polyethylene has the heat sealability of the latter as well as the special barrier and mechanical properties of each component;
(ii) In extrusion blow moulding a polystyrene central layer between two layers of HDPE will confer much enhanced stiffness to the bottle. The polystyrene flow is intermittently interrupted so that this polymer is only found in the body of the bottle, but not near the bottom and the shoulder where the polyethylene is clamped together to make a good heat seal; the end trimmings, being one polymer only, are recovered and recycled.
(iii) Multilayer extrusion introduces an altogether novel concept of multiple lamination in a single-step machine process. It can be taken further by using rotating dies whereby the individual layers can be made as thin as several hundred Ångstrom units. Going even further, one can bring about the break-up of layers into a quasi-fibrillated film which then constitutes a fibrous reinforcement in a continuous matrix of the major component. Eventually the fibrillar continuity can even be further broken up to form a longitudinally orientated streamline of droplets. Such structures are likely to have most interesting and unique properties which neither component can provide on its own; their investigation has barely begun, but it is certain that it holds out most exciting prospects for the future.

9.6. Benefits of Die Design

What, *in summa*, are the benefits of die design?

The principal benefit lies in the combined advantages of greatly reducing the number of test runs which characterises an exclusively empirical approach. Die design approaches an engineering design problem scientifically, using the basic principles of fluid mechanics along with sensible flow models in relatively simple channels such as dies. This results in predictions which are of acceptable accuracy, at least as a first approximation, with an error of no more than about 20%. The use of suitably derived equations is simple and saves much material, machine wear, die redesign costs and overheads, not to mention time. A sound calculation establishes the fundamental variables of the extrusion process which, with a minimum of adjustment through minor modifications, leads to a quick, elegant and inexpensive solution of many problems.

REFERENCES

1. LENK, R. S., *Polymer Rheology*, Applied Science Publishers, London, 1978, Chapter 1.
2. COGSWELL, F. N., and LAMB, P., *Trans. J. Plast. Inst.*, **35**, December 1967, 120.
3. PLAJER, O., *Plastverarbeiter*, **23** (6), 1972, 407.
4. LENK, R. S. and FRENKEL, R. A., *J. Appl. Polym. Sci.*, **26** (8), 1981, 2801.
5. RAMSTEINER, F., *Kunststoffe*, **62** (11), 1972, 766.
6. LENK, R. S., *J. Appl. Polym. Sci.*, **26** (10), 1981, 3939.
7. COGSWELL, F. N., private communication.
8. LENK, R. S., *J. Appl. Polym. Sci.*, **26**(9), 1981, 3171.
9. OH, K. H. and LENK, R. S., *J. Polym. Sci.* (*Polym. Lett. Ed.*), **20**, 1982, 135.
10. OH, K. H., unpublished work.
11. RAMSTEINER, F., *Kunststoffe*, **61**(12), 1971, 943.
12. BAGLEY, E. B. and BIRKS, A. M., *J. Appl. Phys.*, **31**, 1960, 556.
13. CARLEY, J. F., *SPE Journal*, **19**(9), September 1963, 977.
14. CARLEY, J. F., *SPE Journal*, **19**(12), December 1963, 1263.
15. LENK, R. S., *Plastics Rheology*, Maclaren, London, 1972.
16. LENK, R. S. and SLATER, G., *J. Appl. Polym. Sci.*, **23**, 1979, 2793.
17. SISSON, W. E. and MOOREHEAD, F. F., *Textile Res. J.*, **23**, 1953, 152; **30**, 1960, 153.
18. HICKS, E. M., PIPPEPPS, E. A., HEWEPP, J. and BRAND, R. H., *in Man-made Fibres*, H. F. Mark, S. M. Atlas and E. Cernia, Eds. Wiley Interscience, New York, 1967, Volume I, p. 375.
19. HICKS, E. M., *et al.*, *Textile Res. J.*, **30**, 1960, 675.
20. NAITOVE, M. H., *Plast. Technol.* **23**(2), 1977, 61.
21. IWAWAKI, A., KOJIMA, K. and SHITARA, Y., *Mod. Plast. Int.*, **7**, 1977, 15.
22. ANON., *Brit. Plast. Rubber*, October 1981, 37.
23. HAN, C. D., *Rheology of Polymer Processing*, Academic Press, New York and London, 1976.
24. HAN, C. D. and KHAN, J., *J. Appl. Polym. Sci.*, **17**, 1973, 1289.
25. LEE, B. L. and WHITE, J. L., *Trans. Soc. Rheol.*, **18**, 1974, 467.
26. SOUTHERN, J. H. and BALLMAN, R. L., *Appl. Polym. Symp. No. 20*, 1973, 1234.
27. SOUTHERN, J. H. and BALLMAN, R. L., *J. Polym. Sci.*, *A2*, **13**, 1975, 863.
28. HAN, C. D. and CHIN, H. B., *Polym. Eng. Sci.*, **19**(16), December 1979, 1156.
29. SCHRENK, W. J. and ALFREY, T., Jr, *SPE Journal* **29**, June 1973, 134.
30. SCHRENK, W. J. and ALFREY, T., Jr, *SPE Journal*, **29**, July 1973, 43.
31. VAN OENE, H., in *Polymer Blends*, S. Newman and D. R. Paul, Eds., Academic Press, New York, 1978, Volume 1, Chapter 7.

32. WORTH, R. A., *Polym. Eng. Sci*, **20**(8), May 1980, 551.
33. DEBERDEEV, R. Ya., *et al.*, *Int. J. Polym. Mat.*, **3**, 1974, 177.
34. PLOCHOCKI, A., in *Polymer Blends*, S. Newman and D. R. Paul, Eds., Academic Press, 1979, Volume 2, Chapter 21, p. 343.

INDEX

ABS, cellular pipe, 65–6
Accumulator blow moulding, 217–20, 264–5
  accumulator head design, 219
  parison wall thickness controllers, 219
  process details, 217–19
  products, 219–20
  safety features, 219
Acrylic compounds, 173
Acrylonitrile/butadiene/styrene. *See* ABS
Additives in PVC window frame manufacture, 172–4
Air flow rates, 110
Alarm
  annunciation, 39
  logic module, 42
Alarm/protection philosophies, 25
Aluminium window frames, 161
  combined with PVC, 162
Automation in extrusion blow moulding, 216

Bags, 81
Barrel/die zone
  control, 22–5
    block diagram, 25–6
  controller specification, 27–30
Basic Electronics Package, 40
Beta-ray gauge, 101
Biaxial orientation, 226
Blends, blown film production, 148
Blister packing, 82
Blocking effects, 93
Blow moulding processes, 195–227
  development, 195–6
  die design (rheology), 256–8
  market areas, 197
  material consumption levels, 196
  materials used, 196
  see also Accumulator blow moulding; Co-extrusion blow moulding; Extrusion blow moulding; Injection blow moulding; Stretch blow moulding
Blown film production, 75–153
  adhesion between layers, 150
  alternatives to winding, 115
  applications, 79–82
  automatic process control, 115–19
  blends, 148
  co-extrusion, 149–51
  combinations of plastics, 148–51
  commercial limitations, 120–1
  comparison with flat film process, 76
  extruders, 83–4
  film
    formation, 121
    handling, 114–15
    movement in reel, 97
    requirements, 77
    strength properties, 127–9
    structure, 125
    surface treatment, 115

Blown film production—*contd.*
film—*contd.*
thickness
control, 100–1
limitations, 102
history and development, 75–7
lower cost control systems, 119
machine direction extension ratio, 121–2
manufacturing process, 83–102
market
structure, 78–82
volume, 78–9
materials, 78, 131–51
maximum output figures, 102
melt generation, 102–4
microprocessor process control, 115–19
operating control, 100–1
optical properties, 129–31
process
control, 98–101
development, 77, 102–19
quality control, 101
relation between process and properties, 119–31
rheological behaviour, 125
setting film width and thickness, 99
transverse direction extension ratio, 121–2
unevenness in film thickness, 96
Blow-up ratio (BUR), 122, 125, 126, 128, 129, 144
Bottles, 196, 197, 224, 226
Bow roll, 95–6
Bubble
collapse, 114–15
dimensions, 123
guides, 90–3
shape prediction, 125–6
sizing, 112
temperature, 110
Burst firing, 15

Cable insulation
cellular thermoplastics extrusion, 67–9
Cable insulation—*contd.*
chemical expansion systems, 67
co-extrusion techniques, 69
physical expansion systems, 68
polymer and process developments, 68–9
Calcium carbonate filler, 173
Capillary rheometry, 107
Carbon dioxide cooling, 213
Cascade control, 17–18
Cellular thermoplastics extrusion, 45–73
cable insulation, 67–9
chemical expansion system, 67–8
co-extrusion techniques, 56, 57, 69
controlled expansion techniques, 55–6
expansion systems, 51
chemical, 51
physical, 51
free expansion techniques, 54–5, 57
fundamentals, 49–56
hollow profiles and pipe, 64–9
nucleation
factor, 51–3
systems, 53–4
pattern of production and development, 45
physical expansion systems, 68
processing conditions, 49–51
profile production, 57–62
range of processes, 56–69
sheet production, 62–4
specialist processes, 69–72
techniques, 53–6
tubular blown film, 66–7
Celuka controlled expansion technique, 57
Circular channels, 238
'swan neck' type convergence, with, 260–3
Closed loop system, 18
Co-extrusion, 263–71
applications, 263
blow moulding, 195, 222–3
container production, 223
reasons for development, 223
blown film production, 149–51

Co-extrusion—*contd.*
  cable insulation, 69
  cellular thermoplastics extrusion, 56, 57, 69
  equipment, 264
  multilayer flow analysis, 269–71
  pipe production by, 265–6
  reasons for, 277
  rheology of, 267–9
  rotating dies, 273
Cold junction compensation, 14
Collapsing frames, 90–3
Colour
  fastness, 167, 170
  range, 174
Consistency constant, 108
Control
  considerations, 7–22
  equipment, 13–17
  loop monitoring, 39
Controller outputs, 15
Cooling
  rate, 110–11
  rings, 88–9
    adjustments, 100
    dual lip, 111
  stages, 24
  systems, 7, 110
    external, 110–12
    internal, 112–14
    Reinfenhauser design, 113
Creasing, 90–1, 95
Crystallisation, 124–5, 130

Deflashing equipment, 217
Die
  construction, 104
  design
    analysis results, 109
    benefits of, 277
    blown film, 84
    co-extrusion, 60
    criteria for evaluating, 107
    exit gap, 88
    feedback dies, 264
    gap width variation, 106
    LLDPE, for, 141

Die—*contd.*
  design—*contd.*
    materials, 182
    multi-manifold internal combining dies, 264
    plenum chamber feeds and concentricity, 266
    PVC window frames, for, 180
    side fed, 84
    spiral mandrel, 105–7, 109, 123
    star fed, 104
    tubular blown film, 66
  design (rheology), 229
    accuracy of die calculations, 276
    blow moulding, 256–8
    calculation examples, 248
    convergent circular, 244
    convergent rectangular, 244
    convergent square, 244
    converging dies, 276
    cylindrical dies, 243
    design solution, 261–3
    divided annular, output, 253
    elastic effects, 275
    elongational flows, 275
    end corrections, 258–9
    entrance
      corrections, 258–9
      criteria, 260–1
    equilateral triangular converging, 246
    equilateral triangular prismatic, 246
    length/radius function, 261
    linearly tapering, 247
      zero taper angle, with, 247
    parallel-sided, 246
    rectangular converging, 245
    rectangular prismatic, 244
    regular hexagonal prismatic, 250
    regular polygonal converging, 245
    regular polygonal prismatic, 245
    rotating dies, 271–3
    secondary flows, 276
    square prism, 244
    tapered dies, 235
    tapering regular hexagonal, 251

Die—*contd.*
design (rheology)—*contd.*
truncated cone, 248, 249, 252
land length, 252
mandrel, with, 254
wide slit dies, 260
see also Die design *and under specific channel configurations*
Direct digital controller, 34
Discrete controller implementation, 25
Dual control module, 41
Durapipe co-extrusion die, 266

Elliptical channels, 242–3
convergence, without, 246
linear convergence, with, 246
End corrections, 258–9
Entrance
corrections, 258–9
criteria, 260–1
Ethylene propylene diene monomer (EPDM) rubbers, 191
Ethylene/vinyl acetate graft copolymers, 142–3, 170, 171
blends, 148
co-extrusion, 149
Ethylene/vinyl acetate impact modified PVC copolymer, 176
Extruders
autothermal, 103
blown film manufacture, 83–4
screw design, 103
output, and, 83–4
twin-screw, 180
Extrusion
blow moulding, 199–217, 264–5
automation, 216
base and handle design, 212
blow pin design and cooling, 213
calibration system, 211
deflashing, 210
extruder screws and barrels, 199–200
extrusion heads, 202–3
internal cooling, 213–15

Extrusion—*contd.*
blow moulding—*contd.*
length control, 209
machine
control systems, 210–11
movement, 209
mould(s), 211–16
cooling, 211–12
materials, 212
positioning, 209–10
parison
control, 206–9
cutting, 203–6
forming die, 256–8
programming, 207
wall thickness control systems, 207
production line, 217
defects, 131
process
block diagram, 1
control of die inlet condition, 3
controlled or deterministic input variables, 1
PVC window frame manufacture, for, 179–83
uncontrollable or random input variables, 2

Fast cycling, 15
Fibre spinning dies, 273
Fillers in PVC window frame manufacture, 173
Finishing processes, 188–9
Flat film process, 76
Flow
analysis, 108
multilayer extrusion, for, 269
die, through. *See* Die design (rheology)
equations
experimental evaluation of, 248
summary of, 243–7
index, 108
phenotypes, 232
properties, 107
rates, 109

Flow—*contd.*
  steady-state shear-viscous, 235–58
  velocity, 233
Food packaging, 146
Form fill and seal machines, 82
Friction effect, 92

Gain coefficients, 3
Gas mixtures, 53
Geometric effect, 91
Gloenco system, 119
Gloss, 129
Grooved barrels, 103
Gussets, 93

Hagen–Poiseuille equation, 230, 231, 233, 234, 235, 255, 257, 276
Haul-off, 90
Haze, 129, 130
Heat
  ageing stability, 165
  power feedback, 24
  removal, 112
  reversion limits, 165
  stability test, 166
Heating stage, 23
Heating systems, 6
Heat-sensitive materials, 264
Hercules–Bekum process, 227
High density polyethylene (HPDE), 134–6
  bags, 81–2
  blends, 148
  blown film production, 78–9, 83, 121
  co-extrusion, 149, 268
  film
    production, 135
    strength properties, 128–9
  manufacture and structure, 134
  molecular structure effects, 136
  molecular weight, 135
  properties and applications, 135
High molecular weight polyethylene (HMWPE), 219
Hollow profile systems, 59
Homogeneity effects, 2, 3
Hooke's Law, 232
Hunkar ISD system, 214–15

Impact
  modification, 170–2, 172
  resistance at low temperatures, 166
Injection
  blow moulding, 196, 220–2
    advantages, 221–2
    basic process, 220–1
    disadvantages or restrictions, 222
    materials and products, 220
  stretch blow moulding, 226
Integrated Dimensional Control System (IDC), 114, 116–17
Ionomer resins, 143–4
  processing, 144
  structure and properties, 143–4
IR thickness gauge, 117, 118, 151

L/D ratio, 83
Light emitting diodes (LEDs), 207
Linear flow velocity, 258
Linear low density polyethylene (LLDPE), 78, 136–42
  blends, 148
  blown film, 140–2
  film properties, 137–40
  growing utilisation of, 152
  heat sealing range, 142
  manufacture and structure, 136–7
  mechanical properties, 137–40
  melt rheology, 137
  plant requirements, 141
Low density polyethylene (LDPE), 131–4
  bags, 81–2
  blends, 148
  blown film production, 78–9, 83, 84, 121
  cellular crosslinked, 69
  co-extrusion, 149, 267
  film properties, 127, 132–4
  manufacture and structure, 131
  melt flow properties, 132

Low density polyethylene (LDPE) —*contd.*
molecular weight, 132
optical properties, 129–31
process developments, 71
sacks, 81
shrink wrapping, 80
Lubricants in PVC window frame manufacture, 173

Marketing
blow mouldings, 197
PVC window frames, of, 193
stretch blow moulding, 226
Master module, 40
Medium bandwidth system, 20
Melt
cascade
controller, 27
loop, 27
characterisation, 107
extension measurement, 122–4
flow index, 66, 79, 81, 127, 132
pressure, 3, 22, 50, 51, 53, 55, 61, 64
temperature, 3, 22, 49, 51, 55, 131, 142
control, 25
Metalised nitride oxide semiconductor (MNOS), 41, 42
Microprocessor process control, 34–42
alarm annunciation, 39
basic electronics package, 40
blown film manufacture, 115–19
control loop monitoring, 39
direct digital controller, 34
fundamental approaches, 34
implementation, 27, 30–2
industrial solution, 36
operator station, 36, 37
parameter monitoring, 39
sequence controls, 210–11
supervisory scheme, 35
Modelling, 18–22
Molecular orientation angle, 273
Moulds, extrusion blow moulding, 211–16
Multilayer extrusion flow analysis for, 269
Multi-microlayers, 273

Narrow bandwidth system, 20
NC–E system, 117–18
Newtonian fluids, 231–2
Newtonian shear rate, 234
Newton's Law, 232, 233, 234
Nitrile rubber/poly(vinyl chloride) blends, 71
Nitrogen cooling, 213–14
Nucleation in production of cellular extrudates, 51–4
Nylon plastics, blown film, 147–8

On/off control, 7–8
Optical properties
blown film, 129–31
processing conditions, effect of, 130
Organoleptic considerations, 146

Parallel-sided channel(s), 236–7
elliptical, 243
Partially incompatible additives, 53
Particulate crystalline fillers, 53
Phase angle, 15
Phillips–Orbit process, 227
Pigments in PVC window frame manufacture, 173
Pipe production by co-extrusion, 265–6
Poly(acrylate ester) graft copolymers, 172
Polyamides, blown film, 147–8
Polychloroprene, 191
Polyester glass reinforced pipes, 273
Polyethylene, modified, 142–4
Poly(ethylene terephthalate) (PET)
one-step process, 225
stretch blow moulding, 224–5
two-step process, 224–5

Poly(methyl methacrylate) (PMMA), 174, 272
Polyolefins
  cellular pipe, 66
  cellular profiles, 62
  cellular sheet production, 63–4
  specialist processes, 69
Polypropylene
  blown film, 145
  stretch blow moulding, 226–7
Polystyrene
  cellular sheet production, 64
  co-extrusion, 267, 268
  profiles, 61–2
Poly(vinyl chloride), 145–7
  blown film manufacture, 146–7
  bubble shapes, 147
  cellular pipe
    equipment, 64
    formulation, 65–6
  cellular sheet production, 62
  characteristics and applications, 145–6
  hollow profiles and pipe, 64
  melt formulation, 58–9
  processing equipment, 59
  profiles, 57–61
  shrink
    applications, 147
    wrapping, 80
  stretch blow moulding, 225–6
  see also PVC window frame manufacture
Power
  feedback, 24
  index, 236
  Law, 107, 232, 234, 236, 238, 242, 243, 257, 269, 276
    flow through cylindrical channels, 233–4
  supply module, 41
Pressure
  drop, 235, 237, 249–52, 255
  gradient reduction factor (PGRF), 267
Process control, 32–42
  basic problem, 32–4
Processing aids, 173
Proportional control, 8–10
Proportional plus derivative control, 10–11
Proportional plus integral plus derivative control, 11–13
PVC window frame manufacture, 155–94
  ancillary equipment, 182
  approving organisations, 163–4
  availability of British designs, 157
  calibration section, 181
  comparison
    aluminium, with, 161
    steel, with, 161–2
    timber, with, 160–1
  compound
    formulations, 174–5
    properties, 176–9
  co-ordination, 193
  design details, 163
  durability requirements, 160
  extrusion process, 179–83
  fabrication process, 183–93
  finishing, 188–9
  glazing, 190–1
  habitability requirements, 159
  high speed mixing, 175–6
  hot plate welding, 186
  installation process, 192–3
  joining of profiles, 185–90
  machining and assembling, 184–5
  market
    growth, 156
    share, 160
    size and structure, 156–8
  marketing, 193
  material(s), 160–3
    combinations, 162–3
    requirements, 169–74
  output rates, 183–4
  performance
    considerations, 158–63
    requirements, 159
  profile
    design parameters, 178–9
    properties, 165
  proposed UK standard, 164–7
  quality control, 168, 182–3, 193

PVC window frame manufacture—*contd.*
  replacement windows, 157–8
  safety requirements, 159
  sawing, 184
  solvent welding, 189
  specifications, 163–8
  storage, 184
  types of window, 158–9
  weatherseals or weather stripping, 191–2
  weld
    cycles, 187
    strength, 167, 187
  welding operation, 186–7
  window assembly, 190

Quad monitor module, 42
Quality control, 168
  blown film production, in, 101
  PVC window frames, in, 168, 182–3, 193

Rabinowitsch Correction, 234–6, 238 239, 243, 244–6, 276
Radial melt distribution, 218
Recipe, 40
  module, 42
Rectangular channels, 236–8
Regular polygonal channels, 238–40
Relays, 15
Residual strain, 124
Restrictor grid die, 61
Rheology. *See* Die design (rheology)
Rosenheim Institute for Window Technology, 163
Rotating dies, 271–3
  co-extrusion, 273
  combined pressure and drag flow, 271
  rotating mandrels, 271–3

Sacks, 81
Screw design. *See* Extruders
Sensors, 4–6
Shear
  flow through die, 229
  viscosity, 231
Shear-thickening, 232
Shear-thinning, 232
Shift factors, 255
Shrink wrapping, 80
Shrinkage
  allowance, 117
  properties, 126–7
Skin packaging, 82
Slat roll, 95
Slitting, 93–4
Solvent welding, 189
Spiral mandrel
  unrolled, 108
  see also Die design
Spreader roll, 95
Spreading, 95
Stabiliser systems, 172
Start-up procedures, 99
Static mixers, 104
Steel window frames, 161–2
  combined with PVC, 162
Strain
  rates, 124
  relaxation, 122
Stretch
  blow moulding, 196, 223–7
    market application, 226
  wrapping, 80
Styrene-based polymers. *See* Polystyrene
Styrene–butadiene block copolymers, 148
Suddeutsches Kunststoffe Zentrum (SKZ), 163
Swinging roll design, 94

Tapering rectangular channels, 237–8
Temperature
  control, 4, 22–32
  measurement, 4
  selection for co-extrusion processes, 150–1
Thermocouple location, 5–6
  combined deep/shallow arrangement, 6

Thickness detector, 101
Thyristors, 15, 23
Timber window frames, 160–1
Toxicological considerations, 146
Triacs, 15
Triangular channels, 240–2

Undercut containers, 215–16
Unevenness in film thickness, 96

Velocity
  gradient, 233
  profile, 233
Vicat softening point, 143
Vinyl acetate, 142
Viscosity/shear stress plots, 268
Viscosity/temperature functions, 255
Volume flow rate, 258

Weather resistance testing conditions, 166
Weathering behaviour, 167, 169–70
Weatherseals, 191–2
Weatherstripping, 191–2
Web
  guiding, 94
  transport, 93–6
Welding
  machines, 186
  PVC window frames, of, 186
  solvent, 189
Wide bandwidth system, 21
Wide neck containers, 215–16
Winding, 96–8
  centre, 98
  constant tension, 98
  programmed tension, 98
  surface, 97
  variable tension, 98
Window frames. *See* PVC window frame manufacture

Young's modulus, 232

GPSR Compliance

The European Union's (EU) General Product Safety Regulation (GPSR) is a set of rules that requires consumer products to be safe and our obligations to ensure this.

If you have any concerns about our products, you can contact us on ProductSafety@springernature.com

In case Publisher is established outside the EU, the EU authorized representative is:

Springer Nature Customer Service Center GmbH
Europaplatz 3
69115 Heidelberg, Germany

Batch number: 10584446

Printed by Printforce, the Netherlands